FUNCTIONS OF MATHEMATICAL PHYSICS

The New University Mathematics Series

Editor:
Professor E. T. Davies
Department of Mathematics, University of Southampton

This series is intended for readers whose main interest is in mathematics, or who need the methods of mathematics in the study of science and technology. Some of the books will provide a sound treatment of topics essential in any mathematical training, while other, more advanced, volumes will be suitable as preliminary reading for research in the field covered. New titles will be added from time to time.

Brown and Page: *Elements of Functional Analysis*
Burgess: *Analytical Topology*
Cooper: *Functions of a Real Variable*
Curle and Davies: *Modern Fluid Dynamics* (Vols. 1 and 2)
Porteous: *Topological Geometry*
Roach: *Green's Functions: Introductory Theory with Applications*
Rund: *The Hamilton-Jacobi Theory in the Calculus of Variations*
Smith: *Introduction to the Theory of Partial Differential Equations*
Smith: *Laplace Transform Theory*
Spain: *Ordinary Differential Equations*
Spain: *Vector Analysis*
Spain and Smith: *Functions of Mathematical Physics*
Zamansky: *Linear Algebra and Analysis*

Functions of Mathematical Physics

BARRY SPAIN

Head of Department of Mathematics,
Sir John Cass College, London

AND

M. G. SMITH

Principal Lecturer, Department of Mathematics,
Sir John Cass College, London

VAN NOSTRAND REINHOLD COMPANY

LONDON

NEW YORK TORONTO MELBOURNE

VAN NOSTRAND REINHOLD COMPANY LTD
Windsor House, 46 Victoria Street, London S.W.1

INTERNATIONAL OFFICES
New York Toronto Melbourne

Library of Congress Catalog Card No. 61–19849.

First Published 1970

Printed in Great Britain by
Butler & Tanner Ltd, Frome and London

Preface

This book is designed to present a rigorous account of the theory of those functions which occur frequently in the study of physics and other sciences. For each function, the theory is developed systematically from the particular definition which has been chosen.

Prerequisites for the understanding of the text are Cauchy's theorem and its immediate consequences together with some elementary results on power series and ordinary differential equations. All the results needed in the first two categories can be found, for example, in [5], and in the last category in [7].

Chapter 1 contains the existence theorems and techniques for the solution by series of homogeneous linear differential equations of the second order at ordinary and regular points. This is followed in Chapter 2 by a thorough discussion of the contour integral solutions. Chapter 3 is devoted to the oscillation theorems and Sturm–Liouville theory, and these preliminaries are completed in Chapter 4 on asymptotics. The remaining chapters deal individually with the gamma, hypergeometric, confluent hypergeometric, Legendre, Bessel, Laguerre and Hermite functions. No attempt has been made in this book to discuss the theta functions or elliptic functions, which are treated in [18]. Those functions which can be derived as solutions of ordinary differential equations with periodic coefficients have also been omitted, and a discussion of these can be found in [1].

It was impossible in the limited space at our disposal to exhibit all the known properties of these functions. We believe, however, that we have presented all those properties which are generally required by physicists, engineers, etc. We also believe that it will not be too difficult for the conscientious reader to deduce other properties which are not listed. This text should meet the requirements of third-year undergraduates specializing in mathematics, and postgraduate students in physics, chemistry or engineering who have the prerequisites mentioned above.

We realize that the treatment of completeness of orthonormal sets is somewhat less than what is required. However, we have indicated this insufficiency and given references to a fuller treatment.

In addition to the bibliography at the end of the text, we draw the reader's attention to

ERDELYI, A., MAGNUS, W., OBERHETTINGER, F. and TRICOMI, F. G. *Higher Transcendental Functions*, Vols. I, II, III, McGraw-Hill, New York (1953–55).

MAGNUS, W. and OBERHETTINGER, F. *Formulae and Theorems for the Functions of Mathematical Physics*, Springer, Berlin (1954).

These books contain an encyclopedic summary of the properties of functions, but the explanation is minimal.

Further, numerical data and graphs on some of the functions can be found in

ABRAMOWITZ, M. and STEGUN, I. A. *Handbook of Mathematical Functions*, National Bureau of Standards, Washington, D.C. (1965).

JAHNKE, E. and EMDE, F. *Tables of Functions*, Dover, New York (1960).

The authors would like to thank Mrs. M. G. Smith, who typed the manuscript and prepared the index. This book was planned and developed from lecture courses given at Sir John Cass College. We wish to thank those students whose contributions in discussion helped shape the final version.

BARRY SPAIN
M. G. SMITH

Sir John Cass College
July 1968

Contents

CHAPTER 1

Series Solution of Second-order Linear Homogeneous Equations

1.1 Ordinary and singular points

This text is concerned primarily with certain functions which are regular and satisfy a homogeneous linear differential equation of the second order. Consequently we investigate such differential equations and examine the effect which the singularities of the coefficients have on their solutions.

The second-order homogeneous linear differential equation can always be displayed in the form

$$L(w) \equiv \frac{d^2w}{dz^2} + p(z)\frac{dw}{dz} + q(z)w = 0, \tag{1.1.1}$$

where $p(z)$ and $q(z)$ are functions of the complex variable z.

We say that z_0 is an ordinary point of the differential equation if $p(z)$ and $q(z)$ are regular functions of z in some neighbourhood of z_0. In this case Taylor expansions

$$\sum_0^\infty p_s(z - z_0)^s \text{ and } \sum_0^\infty q_s(z - z_0)^s$$

exist and converge within some circle with centre at z_0.

Points which are not ordinary are called singular. At a singular point z_0, either one or both of the functions $p(z)$ and $q(z)$ have a singularity. Singular points are divided into two kinds. The point z_0 is called a regular singularity or regular point or a singular point of the first kind if $(z - z_0)p(z)$ and $(z - z_0)^2q(z)$ are regular functions of z in some neighbourhood of z_0.

A point which is neither ordinary nor regular is called an irregular singularity or an irregular point of the differential equation.

The change of the independent variable from z to Z by $z = Z + z_0$ transfers the finite point $z = z_0$ to the origin $Z = 0$ without essentially changing the form of the differential equation. Accordingly, there is no loss in generality if we restrict our investigation to the solution of differential equations in the neighbourhood of the origin. A different substitution will be introduced later to discuss the nature of the solution at infinity.

1.2 Solution near an ordinary point

We begin by looking for a formal solution of (1.1.1) in the form of the power series

$$w(z) = \sum_0^\infty c_s z^s.$$

Let the Taylor expansions of the regular functions $p(z)$ and $q(z)$ be

$$p(z) = \sum_0^\infty p_n z^n, \quad q(z) = \sum_0^\infty q_n z^n$$

and let the smaller of the two radii of convergence be R.

Formal substitution in the differential equation yields

$$\sum_2^\infty s(s-1)c_s z^{s-2} + \sum_0^\infty p_n z^n \sum_1^\infty s c_s z^{s-1} + \sum_0^\infty q_n z^n \sum_0^\infty c_s z^s \equiv 0,$$

from which

$$(s+2)(s+1)c_{s+2} + \sum_{l=0}^{s} \{(l+1)p_{s-l}c_{l+1} + q_{s-l}c_l\} = 0 \qquad (1.2.1)$$

for $s = 0, 1, 2, \ldots$

These equations, in which the coefficient of c_{s+2} never vanishes, are linear, and so by successive calculations $c_2, c_3, \ldots, c_s, \ldots$ can be determined as linear combinations of c_0 and c_1 in the form

$$c_s = \lambda_s c_0 + \mu_s c_1 \qquad (s = 2, 3, \ldots).$$

In this way the formal solution

$$w(z) = \sum_0^\infty c_s z^s \equiv \sum_0^\infty (\lambda_s c_0 + \mu_s c_1) z^s, \qquad (1.2.2)$$

in which $\lambda_0 = \mu_1 = 1$, $\lambda_1 = \mu_0 = 0$, has been obtained.

At this stage there arises the question of the convergence of the series (1.2.2). The proof of convergence which follows relies on setting up a convergent series which by comparison makes (1.2.2) convergent.

Since the series for $p(z)$ and $q(z)$ converge for $|z| < R$, there exists a constant $M > 0$ such that

$$|p_n| R^n \leqslant M, \qquad |q_n| R^n \leqslant M$$

for all integral values of n. Hence it follows from (1.2.1) that

$$(s+2)(s+1)|c_{s+2}| \leqslant (M/R^s) \sum_{l=0}^{s} \{(l+1)|c_{l+1}| + |c_l|\}R^l + M|c_{s+1}|R, \tag{1.2.3}$$

where the term $M|c_{s+1}|R$ has been included for a reason which will become clear at a later stage.

Now *define* positive constants C_s by the equations

$$C_0 = |c_0|, \qquad C_1 = |c_1|,$$

$$(s+2)(s+1)C_{s+2} = (M/R^s) \sum_{l=0}^{s} \{(l+1)C_{l+1} + C_l\}R^l + MC_{s+1}R. \tag{1.2.4}$$

From equations (1.2.3) and (1.2.4) we easily deduce that $C_2 \geqslant |c_2|$ and that the assumption $C_l \geqslant |c_l|$ for $l = 2, 3, \ldots, s+1$ implies that $C_{s+2} \geqslant |c_{s+2}|$. Hence by the process of induction

$$C_s \geqslant |c_s| \geqslant 0$$

for all integral s.

Replace s by $s - 1$ in (1.2.4) and the result is

$$(s+1)sC_{s+1} = (M/R^{s-1}) \sum_{l=0}^{s-1} \{(l+1)C_{l+1} + C_l\}R^l + MC_sR.$$

Hence we have

$$(s+2)(s+1)C_{s+2} - (s+1)sC_{s+1}/R = (M/R^s)\{(s+1)C_{s+1} + C_s\}R^s + MC_{s+1}R - MC_s.$$

Now we see that the introduction of $MC_{s+1}R$ in (1.2.4) ensures that a two-term recurrence relation is obtained. On simplification we have

$$\frac{C_{s+2}}{C_{s+1}} = \frac{(s+1)s + M(s+1)R + MR^2}{(s+2)(s+1)R},$$

and so $\lim_{s\to\infty} C_{s+2}/C_{s+1} = 1/R$. Accordingly, the ratio test establishes that the series $\sum_0^\infty C_s z^s$ converges in $|z| < R$, and so by the comparison test $\sum_0^\infty c_s z^s$ converges absolutely in $|z| < R$.

The formal processes of term-by-term differentiation, multiplication and regrouping of power series by which $w(z) = \sum_{0}^{\infty} c_s z^s$ was made to satisfy the differential equation are now justified, since all the series converge uniformly and absolutely in every closed domain within $|z| = R$.

Further, we see that $c_0 = w(0)$ and $c_1 = w'(0)$, and so we have established the general result.

The differential equation $w'' + p(z)w' + q(z)w = 0$, where $p(z)$ and $q(z)$ are regular in some neighbourhood of the ordinary point z_0 has a unique solution $w(z)$ which satisfies the initial conditions $w(z_0) = a$ and $w'(z_0) = b$, where a and b are arbitrary constants. Further, this solution is regular in $|z| < R$, where R is the smaller of the radii of convergence of the Taylor expansions of $p(z)$ and $q(z)$ about z_0.

Corresponding to the choices $c_1 = 0$ and $c_0 = 0$ we have established two solutions $w_1(z)$ and $w_2(z)$ given by

$$w_1(z) = 1 + \sum_{s=2}^{\infty} \lambda_s z^s, \qquad w_2(z) = z + \sum_{s=2}^{\infty} \mu_s z^s.$$

It is obvious that $w_1(z)$ and $w_2(z)$ are linearly independent.

The functions $w_1(z)$ and $w_2(z)$ are defined in the first place in the neighbourhood of the origin. The process of analytic continuation can be carried out along any path which does not pass through a singular point of the differential equation, and so the solution can be defined wherever $p(z)$ and $q(z)$ are both regular. However, the continuation may not be single-valued if the differential equation has singular points in a finite domain.

ILLUSTRATION 1

The origin is an ordinary point of Legendre's differential equation

$$(1 - z^2)w'' - 2zw' + \nu(\nu + 1)w = 0$$

and so there exist two regular solutions in $|z| < 1$.

Calculation shows that (1.2.1) reduces to

$$(s + 2)(s + 1)c_{s+2} = (s - \nu)(s + \nu + 1)c_s$$

and so we arrive at the solutions

$$w_1(z) = 1 - \frac{\nu(\nu + 1)}{2!}z^2 + \frac{\nu(\nu - 2)(\nu + 1)(\nu + 3)}{4!}z^4 - \ldots,$$

$$w_2(z) = z - \frac{(\nu - 1)(\nu + 2)}{3!}z^3 + \frac{(\nu - 1)(\nu - 3)(\nu + 2)(\nu + 4)}{5!}z^5 - \ldots.$$

By using the gamma function (for notation see section 5.1 and in particular (5.1.3)) $\Gamma(z)$ which possesses the property $\Gamma(z) = (z-1)\,\Gamma(z-1)$ we can express these solutions in the forms

$$w_1(z) = \frac{\Gamma(\frac{1}{2}\nu + 1)}{\Gamma(\frac{1}{2}\nu + \frac{1}{2})} \sum_0^\infty (-1)^s \frac{\Gamma(\frac{1}{2}\nu + s + \frac{1}{2})(2z)^{2s}}{\Gamma(\frac{1}{2}\nu - s + 1)(2s)!}, \tag{1.2.5}$$

$$w_2(z) = \frac{\Gamma(\frac{1}{2}\nu + \frac{1}{2})}{2\,\Gamma(\frac{1}{2}\nu + 1)} \sum_0^\infty (-1)^s \frac{\Gamma(\frac{1}{2}\nu + s + 1)(2z)^{2s+1}}{\Gamma(\frac{1}{2}\nu - s + \frac{1}{2})(2s+1)!}. \tag{1.2.6}$$

It follows immediately from the ratio test that the radius of convergence of both $w_1(z)$ and $w_2(z)$ is unity.

We see that $w_1(z)$ is a polynomial of degree n when ν is an even integer n, and $w_2(z)$ is a polynomial of degree n when ν is an odd integer n. These polynomials, multiplied respectively by the constants

$$\frac{(-1)^{n/2} n!}{2^n\{(n/2)!\}^2} \quad \text{and} \quad \frac{(-1)^{(n-1)/2}(n+1)!}{2^n\{(n-1)/2\}!\{(n+1)/2\}!},$$

are called Legendre polynomials and denoted by $P_n(z)$.

ILLUSTRATION 2

The origin is an ordinary point of Hermite's differential equation

$$w'' - 2zw' + 2\nu w = 0$$

and so there exist two regular solutions which are entire functions.

In this case (1.2.1) becomes

$$(s+2)(s+1)c_{s+2} = 2(s-\nu)c_s$$

and so we obtain the solutions

$$w_1(z) = 1 - \frac{2\nu}{2!}z^2 + \frac{2^2\nu(\nu-2)}{4!}z^4 - \ldots,$$

$$w_2(z) = z - \frac{2(\nu-1)}{3!}z^3 + \frac{2^2(\nu-1)(\nu-3)}{5!}z^5 - \ldots.$$

By using the gamma function we may express these solutions in the forms

$$w_1(z) = \Gamma(\tfrac{1}{2}\nu + 1) \sum_0^\infty (-1)^s (2z)^{2s} \Big/ \Big\{(2s)!\,\Gamma(\tfrac{1}{2}\nu - s + 1)\Big\}, \tag{1.2.7}$$

$$w_2(z) = \tfrac{1}{2}\,\Gamma(\tfrac{1}{2}\nu + \tfrac{1}{2}) \sum_0^\infty (-1)^s (2z)^{2s+1} \Big/ \Big\{(2s+1)!\,\Gamma(\tfrac{1}{2}\nu - s + \tfrac{1}{2})\Big\}. \tag{1.2.8}$$

We see that $w_1(z)$ is a polynomial of degree n when ν is an even integer

n, and that $w_2(z)$ is a polynomial of degree n when ν is an odd integer n. These polynomials, multiplied respectively by the constants

$$\frac{(-1)^{n/2}n!}{(n/2)!} \quad \text{and} \quad \frac{2(-1)^{(n-1)/2}n!}{\{(n-1)/2\}!},$$

are called Hermite polynomials and denoted by $H_n(z)$.

1.3 Indicial equation at a regular singularity

When the origin is a regular point of the differential equation (1.1.1), we begin by looking for a formal solution of the type

$$w(z) = z^\alpha \sum_0^\infty c_s z^s, \qquad (c_0 \neq 0). \tag{1.3.1}$$

Since $zp(z)$ and $z^2q(z)$ are regular at the origin, we can expand these functions into their Taylor series

$$zp(z) = \sum_0^\infty p_n z^n, \qquad z^2q(z) = \sum_0^\infty q_n z^n.$$

Let the smaller of the two radii of convergence be R.

Formal substitution in the differential equation yields

$$\sum_0^\infty (s+\alpha)(s+\alpha-1)c_s z^{s+\alpha} + \sum_0^\infty p_n z^n \sum_0^\infty (s+\alpha)c_s z^{s+\alpha} + \sum_0^\infty q_n z^n \sum_0^\infty c_s z^{s+\alpha} \equiv 0.$$

We note that the terms in z^α do not lead to a recurrence relation but to the equation

$$c_0 F(\alpha) = 0, \tag{1.3.2}$$

where

$$F(\alpha) \equiv \alpha(\alpha - 1) + p_0\alpha + q_0.$$

The other terms yield the recurrence relation

$$c_s F(\alpha + s) + \sum_{l=0}^{s-1} c_l\{(l+\alpha)p_{s-l} + q_{s-l}\} = 0 \tag{1.3.3}$$

for $s \geqslant 1$.

Equation (1.3.2) is satisfied by choosing α to be a root of the so-called indicial equation $F(\alpha) = 0$. The roots α_1 and α_2 are called the exponents at the regular singular point.

After choosing a value of α, the set of linear equations (1.3.3) determines c_s as successive multiples of c_0. In this way two solutions correspond in general to the two exponents α_1 and α_2.

When the roots of the indicial equations are equal, this method obviously leads to one and only one formal solution.

This process may break down if at any stage $F(\alpha + s)$ in equation (1.3.3) vanishes for some value of the integer s. It is clear that this can only happen if the roots differ by an integer and if α is chosen to be that root with the smaller real part. Accordingly, when $\alpha_1 - \alpha_2$ is an integer and $\mathscr{R}(\alpha_1) > \mathscr{R}(\alpha_2)$ there exists one formal solution of the type (1.3.1), and this solution corresponds to the exponent α_1.

A detailed discussion of the behaviour of the solution in this case will be taken up in section 1.6.

In the next section we examine the convergence of the series $\sum_0^\infty c_s z^s$ by a method similar to that of the previous section.

It will be shown that $\sum_0^\infty c_s z^s$ is regular in $|z| < R$. However, the regularity of the solution (1.3.1) may be affected at the origin by the presence of the factor z^α. If α is a positive integer or zero the solution will be regular in $|z| < R$, but for any other value of α the solution is only regular in the annulus $0 < |z| < R$ with a suitable cut.

The reader is warned against assuming that the solutions obtained in this way are not convergent for $|z| > R$. Each example must be examined to find out if its radius of convergence is greater than R.

For example, the differential equation

$$(1 - z)zw'' + 2z(1 + z)w' - 2(1 + z)w = 0$$

has a regular singularity at the origin with exponents 0 and 1. We easily verify that $w = z$ is a solution which is an entire function despite the fact that the radius of convergence of the Taylor series is unity for both $zp(z)$ and $z^2q(z)$.

1.4 Solution near a regular singularity when the exponents differ by a non-integer

Let α_1 and α_2 with $\mathscr{R}(\alpha_1) \geqslant \mathscr{R}(\alpha_2)$ be the roots of the indicial equation $F(\alpha) = 0$. That is,

$$F(\alpha) \equiv \alpha(\alpha - 1) + p_0\alpha + q_0 \equiv (\alpha - \alpha_1)(\alpha - \alpha_2).$$

Hence

$$F(\alpha_1 + s) = s(s + \alpha_1 - \alpha_2)$$

and so

$$|F(\alpha_1 + s)| \geqslant s(s - |\alpha_1 - \alpha_2|). \tag{1.4.1}$$

Again there is a constant $M > 0$ such that

$$|p_n| R^n \leqslant M \quad \text{and} \quad |q_n| R^n \leqslant M$$

for integral n. It follows from (1.3.3) and (1.4.1) that

$$s(s - |\alpha_1 - \alpha_2|) |c_s| \leqslant M \sum_{l=0}^{s-1} (l + |\alpha_1| + 1) |c_l| R^{l-s}. \quad (1.4.2)$$

Choose the integer N so that

$$N - 1 \leqslant |\alpha_1 - \alpha_2| < N$$

and define the positive constants C_r by

$$C_r = |c_r| \qquad \text{for } r = 0, 1, 2, \ldots, (N-1)$$

and C_s for $s \geqslant N$ by

$$s(s - |\alpha_1 - \alpha_2|)C_s = M \sum_{l=0}^{s-1} (l + |\alpha_1| + 1)C_l R^{l-s}. \quad (1.4.3)$$

By an inductive proof it follows as in the previous section that

$$C_s \geqslant |c_s| \geqslant 0$$

for all integral s.

The result of combining (1.4.3) with the equation obtained on replacing s by $s - 1$ is, on simplification,

$$\frac{C_s}{C_{s-1}} = \frac{(s-1)(s-1-|\alpha_1 - \alpha_2|) + M(s + |\alpha_1|)}{Rs(s - |\alpha_1 - \alpha_2|)}$$

and so $\lim_{s\to\infty} C_s/C_{s-1} = 1/R$. Accordingly the ratio test shows that $\sum_0^\infty C_s z^s$ converges for $|z| < R$ and so by the comparison test $\sum_0^\infty c_s z^s$ converges absolutely for $|z| < R$.

The presence of the multiplying factor z^{α_1} in (1.3.1) may introduce a branch point at the origin. Thus we can at least state that $z^{\alpha_1} \sum_0^\infty c_s z^s$ is a solution of the differential equation and is regular for $0 < |z| < R$, if the annulus is cut to make the solution single-valued.

Similarly we can show that $z^{\alpha_2} \sum_0^\infty c_s z^s$ is an independent solution regular in the cut annulus $0 < |z| < R$.

Illustration 3

The origin is a regular singular point of Laguerre's differential equation

$$zw'' + (\alpha + 1 - z)w' + \nu w = 0.$$

Since α occurs in this equation, let us use β for the exponents. We readily obtain $\beta_1 = 0$ and $\beta_2 = -\alpha$. We examine here the case where α is neither zero nor an integer. A simple calculation yields the two-term recurrence relation

$$(\beta + s)(\beta + s + \alpha)c_s = (\beta + s - 1 - \nu)c_{s-1}, \qquad s \geqslant 1.$$

Corresponding to $\beta_1 = 0$ and $\beta_2 = -\alpha$ we have

$$s(s + \alpha)c_s = (s - 1 - \nu)c_{s-1}$$

and

$$s(s - \alpha)c_s = (s - 1 - \nu - \alpha)c_{s-1},$$

leading, with suitable choice of the constant c_0, to the solutions

$$w_1(z) = \sum_{s=0}^{\infty} \frac{(-1)^s\,\Gamma(\alpha + 1)\,\Gamma(\nu + 1)\,z^s}{s!\,\Gamma(s + \alpha + 1)\,\Gamma(\nu + 1 - s)} \tag{1.4.4}$$

and

$$w_2(z) = z^{-\alpha} \sum_{s=0}^{\infty} \frac{(-1)^s\,\Gamma(\alpha + \nu + 1)\,\Gamma(1 - \alpha)\,z^s}{s!\,\Gamma(s - \alpha + 1)\,\Gamma(\alpha + \nu + 1 - s)} \tag{1.4.5}$$

respectively.

For non-negative integer values n of ν, $w_1(z)$ reduces to a polynomial of degree n. This polynomial when multiplied by the constant $\{\Gamma(n + \alpha + 1)\}/\{n!\,\Gamma(\alpha + 1)\}$ is called the Laguerre polynomial $L_n^{(\alpha)}(z)$.

Illustration 4

The origin is a regular singular point of Bessel's differential equation

$$z^2w'' + zw' + (z^2 - \nu^2)w = 0, \qquad \mathscr{R}(\nu) \geqslant 0.$$

The exponents at the origin are $\pm 2\nu$ and we restrict ourselves here to the case where ν is not an integer.

A simple calculation yields the equations

$$c_0(\alpha + \nu)(\alpha - \nu) = 0,$$
$$c_1(\alpha + \nu + 1)(\alpha - \nu + 1) = 0,$$
$$c_s(s + \alpha + \nu)(s + \alpha - \nu) = -c_{s-2}, \qquad (s \geqslant 2).$$

We see that $c_1 = 0$ since $2\nu \neq \pm 1$, and so we have

$$s(s + 2\nu)c_s = -c_{s-2}$$

and

$$s(s - 2\nu)c_s = -c_{s-2}$$

corresponding to the respective exponents $+\nu$ and $-\nu$. With suitable choice of the constant c_0 these lead respectively to the solutions

$$w_1(z) = \sum_{s=0}^{\infty} \frac{(-1)^s}{s!\,\Gamma(\nu + s + 1)}\left(\frac{z}{2}\right)^{\nu+2s} \tag{1.4.6}$$

and

$$w_2(z) = \sum_{s=0}^{\infty} \frac{(-1)^s}{s!\,\Gamma(-\nu + s + 1)}\left(\frac{z}{2}\right)^{-\nu+2s}. \tag{1.4.7}$$

These solutions are regular for $|z| > 0$. The function $w_1(z)$ is called the Bessel function of order ν of the first kind and is denoted by $J_\nu(z)$; $w_2(z)$ is the Bessel function of order $-\nu$ and is denoted by $J_{-\nu}(z)$. However, both $w_1(z)$ and $w_2(z)$ exist even in the case when ν is half an odd integer, and so we still have two solutions $J_\nu(z)$ and $J_{-\nu}(z)$ even though 2ν is an odd integer. If ν is a positive integer n, we have

$$\begin{aligned} J_{-n}(z) = w_2(z) &= \sum_{s=0}^{\infty} \frac{(-1)^s}{s!\,\Gamma(-n + s + 1)}\left(\frac{z}{2}\right)^{-n+2s} \\ &= \sum_{s=n}^{\infty} \frac{(-1)^s}{s!\,\Gamma(-n + s + 1)}\left(\frac{z}{2}\right)^{-n+2s} \\ &= \sum_{r=0}^{\infty} \frac{(-1)^{r+n}}{(n + r)!\,\Gamma(r + 1)}\left(\frac{z}{2}\right)^{n+2r} \\ &= (-1)^n J_n(z). \end{aligned}$$

Hence, except for positive integer and zero n, we have found two independent solutions of Bessel's equation.

1.5 Solution near a regular singularity when the exponents are equal

Corresponding to the repeated root $\alpha_1 = \alpha_2$ of the indicial equation $F(\alpha) = 0$ we have

$$F(\alpha_1) = \left(\frac{\partial F}{\partial \alpha}\right)_{\alpha=\alpha_1} = 0,$$

and there exists the solution

$$w_1(z) = z^{\alpha_1} \sum_{0}^{\infty} c_s z^s.$$

We now evaluate $L(w)$ (see (1.1.1)), when α is *not* a solution of the indicial equation but when the coefficients c_s do satisfy the recurrence relation (1.3.3). In virtue of this the coefficients of $z^{\alpha+s}$ vanish for $s \geqslant 1$ and calculation gives

$$L(w) = c_0 z^\alpha F(\alpha).$$

Accordingly, we have

$$L\left(\frac{\partial w}{\partial \alpha}\right) = \frac{\partial}{\partial \alpha} L(u) = c_0 z^\alpha \{F'(\alpha) + F(\alpha) \log z\}$$

and so it follows that

$$L\left\{\left(\frac{\partial w}{\partial \alpha}\right)_{\alpha=\alpha_1}\right\} = 0.$$

That is, $(\partial w/\partial \alpha)_{\alpha=\alpha_1}$ is a formal solution of the differential equation. However,

$$\frac{\partial w}{\partial \alpha} = z^\alpha \log z \sum_0^\infty c_s(\alpha) z^s + z^\alpha \sum_0^\infty c_s'(\alpha) z^s,$$

where the c_s, with the exception of the constant c_0, are at this stage functions of α. Hence, we obtain the formal solution

$$w_2(z) = w_1(z) \log z + z^{\alpha_1} \sum_1^\infty c_s'(\alpha_1) z^s. \tag{1.5.1}$$

Since $\sum_1^\infty c_s(\alpha_1) z^s$ is uniformly and absolutely convergent for $|z| \leqslant R_1 < R$, it follows that $\sum_1^\infty c_s'(\alpha_1) z^s$ is uniformly and absolutely convergent for $|z| \leqslant R_1 < R$. (This justifies our assumption that differentiation with respect to α can be performed term by term.) Consequently the solution $w_2(z)$ is regular for $0 < |z| < R$ provided that a cut is taken to ensure that $\log z$ remains single-valued.

Note carefully that the term in $\log z$ is always present in one of the solutions when the exponents are equal.

ILLUSTRATION 5

From illustration 4 one solution of Bessel's equation of order zero, namely $zw'' + w' + zw = 0$, is

$$w_1(z) = J_0(z) = \sum_0^\infty \frac{(-1)^s}{(s!)^2} \left(\frac{z}{2}\right)^{2s} \tag{1.5.2}$$

and it is an integral function.

The exponents at the origin are both zero, and the recurrence relation becomes

$$c_s(s + \alpha)^2 = -c_{s-2}$$

since $\nu = 0$. With choice $c_0 = 1$ we obtain the solution

$$w(z) = z^{\alpha} \sum_{s=0}^{\infty} (-1)^s z^{2s} / \{(\alpha + 2)^2(\alpha + 4)^2 \ldots (\alpha + 2s)^2\}.$$

That is,

$$c_s(\alpha) = (-1)^s / \{(\alpha + 2)^2(\alpha + 4)^2 \ldots (\alpha + 2s)^2\}$$

and so

$$c_s'(\alpha)/c_s(\alpha) = -2 \sum_{r=1}^{s} \frac{1}{\alpha + 2r}.$$

At this stage we place $\alpha = \alpha_1 = 0$ and we have

$$c_s'(0)/c_s(0) = -\sum_{r=1}^{s} \frac{1}{r}$$

and so the second solution is

$$w_2(z) = Y^{(0)}(z) = J_0(z) \log z - \sum_{1}^{\infty} \frac{(-1)^s}{(s!)^2}\left(1 + \tfrac{1}{2} + \ldots + \frac{1}{s}\right)\left(\frac{z}{2}\right)^{2s}. \tag{1.5.3}$$

$Y^{(0)}(z)$ is the Bessel function of zero order of the second kind and is regular for $|z| > 0$ with appropriate cuts to ensure the single-valuedness of $\log z$.

ILLUSTRATION 6

We consider the solutions of Legendre's equation

$$(1 - z^2)w'' - 2zw' + \nu(\nu + 1)w = 0$$

in the neighbourhood of the regular singularity at $z = 1$. The transformation $z = 1 - 2\xi$ removes the singularity to the origin and the equation becomes

$$\xi(1 - \xi)\ddot{w} + (1 - 2\xi)\dot{w} + \nu(\nu + 1)w = 0$$

where dots denote derivatives with respect to ξ.

The exponents at the origin are both zero, and the recurrence relation

$$c_s(s + \alpha)^2 = c_{s-1}(s + \alpha + \nu)(s + \alpha - \nu - 1), \qquad (s \geqslant 1) \tag{1.5.4}$$

reduces to

$$s^2 c_s = (s + \nu)(s - \nu - 1)c_{s-1}.$$

Accordingly,

$$(s!)^2 c_s = \frac{\Gamma(\nu + s + 1)\,\Gamma(-\nu + s)}{\Gamma(\nu + 1)\,\Gamma(-\nu)} c_0$$

except when ν is an integer; and so we obtain the solution

$$w_1(z) = \frac{1}{\Gamma(-\nu)\,\Gamma(\nu+1)} \sum_0^\infty \frac{\Gamma(-\nu+s)\,\Gamma(\nu+s+1)}{(s!)^2} \left(\frac{1-z}{2}\right)^s. \tag{1.5.5}$$

Since $\lim_{s\to\infty} |c_s/c_{s-1}| = 1$ it follows that this series converges for $|z-1| < 2$ but diverges for $|z-1| > 2$.

When ν is a positive integer n, we have $c_s = 0$ for $s = n+1$ and we verify that

$$(s!)^2 c_s = (-1)^s (s+n)!\,c_0/(n-s)!$$

and so with $c_0 = 1$ we have the polynomial solution of degree n

$$\sum \frac{(-1)^s (n+s)!}{(n-s)!\,(s!)^2} \left(\frac{1-z}{2}\right)^s. \tag{1.5.6}$$

There is another polynomial solution of degree $n-1$ corresponding to the case when ν is a negative integer $-n$.

To obtain the second solution of Legendre's equation at $z = 1$, return to the recurrence relation (1.5.4) whose solution with $c_0 = 1$ is

$$c_s = \prod_{r=1}^s (r+\alpha+\nu) \prod_{r=1}^s (r+\alpha-\nu-1) \Big/ \left\{\prod_{r=1}^s (r+\alpha)\right\}^2.$$

Logarithmic differentiation yields

$$\frac{c_s'(\alpha)}{c_s(\alpha)} = \sum_{r=1}^s \left(\frac{1}{r+\alpha+\nu} + \frac{1}{r+\alpha-\nu-1} - \frac{2}{r+\alpha}\right)$$

and so

$$c_s'(0) = c_s(0) \sum_{r=1}^s \left(\frac{1}{r+\nu} + \frac{1}{r-\nu-1} - \frac{2}{r}\right).$$

Accordingly, the second solution by (1.5.1) is

$$w_1(z)\log z + \frac{1}{\Gamma(-\nu)\,\Gamma(\nu+1)} \sum_{s=1}^\infty \frac{\Gamma(-\nu+s)\,\Gamma(\nu+s+1)}{(s!)^2} \times \left\{\sum_{r=1}^s \left(\frac{1}{r+\nu} + \frac{1}{r-\nu-1} - \frac{2}{r}\right)\right\} \left(\frac{1-z}{2}\right)^s. \tag{1.5.7}$$

1.6 Solution near a regular singularity when the exponents differ by an integer

Let the exponents at the origin be α_1, α_2 where $\mathcal{R}(\alpha_1) > \mathcal{R}(\alpha_2)$ and $\alpha_1 - \alpha_2 = n$, a positive integer. We have seen in section 1.4 that there is a solution $w_1(z) = z^{\alpha_1} \sum c_s z^s$ regular in $0 < |z| < R$.

Corresponding to α_2 we can obtain $c_s(\alpha_2)$ for $s = 1, 2, \ldots, (n-1)$ as linear multiples of c_0 by repeated application of (1.3.3). We find that the coefficient of $c_n(\alpha_2)$ is $F(\alpha_2 + n) = F(\alpha_1) = 0$ and so we cannot obtain a finite value of $c_n(\alpha_2)$. To obviate this difficulty we choose $c_0(\alpha) = \alpha - \alpha_2$ at this stage. It follows that $c_s(\alpha_2) = 0$ for $s = 0, 1, 2, \ldots, (n-1)$, but $c_n(\alpha_2)$ is now indeterminate. Let us choose an arbitrary value of $c_n(\alpha_2)$. Repeated application of (1.3.3) now yields $c_{n+s}(\alpha_2)$ as linear multiples of $c_n(\alpha_2)$ for positive integral s. This process produces the solution

$$z^{\alpha_2} \sum_{s=n}^{\infty} c_s(\alpha_2) z^s = z^{\alpha_1} \sum_{r=0}^{\infty} c_r(\alpha_2) z^r.$$

Since successive coefficients are calculated from (1.3.3), it follows that this solution is merely a constant multiple of the solution $w_1(z)$.

Further progress is now made by evaluating $L(w)$ where $w = z^{\alpha} \sum_{s=0}^{\infty} c_s(\alpha) z^s$ when α is *not* a solution of the indicial equation but the coefficients $c_s(\alpha)$ have been evaluated by the recurrence relation (1.3.3). It follows that the coefficient of $z^{\alpha+s}$ vanishes for $s \geqslant 1$ and the only contribution leads to

$$\begin{aligned} L\{w(z)\} &= (\alpha - \alpha_2) F(\alpha) z^{\alpha} \\ &= (\alpha - \alpha_1)(\alpha - \alpha_2)^2 z^{\alpha}. \end{aligned}$$

On account of the repeated factor we have

$$L\left\{\left(\frac{\partial w}{\partial \alpha}\right)_{\alpha=\alpha_2}\right\} = 0$$

and so $(\partial w/\partial \alpha)_{\alpha=\alpha_2}$ is a solution. However,

$$\frac{\partial w}{\partial \alpha} = z^{\alpha} \log z \sum_{s=0}^{\infty} c_s(\alpha) z^s + z^{\alpha} \sum_{s=0}^{\infty} c_s'(\alpha) z^s.$$

We have seen that the choice $\alpha = \alpha_2$ corresponds to a constant multiple of the solution $w_1(z)$. Consequently, we obtain the formal solution

$$w_2(z) = z^{\alpha_2} \sum_{s=0}^{\infty} c_s'(\alpha_2) z^s + c \log z \, w_1(z), \tag{1.6.1}$$

where c is some constant and the solution $w_2(z)$ is regular for $0 < |z| < R$ with an appropriate cut.

In exceptional cases the constant c may vanish and this is what has happened to Bessel's equation, illustration 4, when 2ν was integral but ν was not an integer. (For another example see illustration 8.) This exceptional case is characterized by the fact that though $F(\alpha + s) = 0$, for $\alpha = \alpha_2$ and $s = \nu$, we also have

$$\sum_{l=0}^{n-1} c_l\{(l + \alpha_2)p_{n-l} + q_{n-l}\} = 0$$

and so the recurrence relation (1.3.3) vanishes identically.

ILLUSTRATION 7

We now obtain the second solution of Bessel's equation

$$z^2w'' + zw' + (z^2 - n^2)w = 0,$$

when n is a positive integer.

The exponents at the origin are $+n$ and $-n$. From illustration 4 we have $c_1 = 0$ and

$$c_s(s + \alpha + n)(s + \alpha - n) = -c_{s-2},$$

from which we have the solution

$$w = z^\alpha \sum_0^\infty (-1)^s(\alpha + n)z^{2s}$$
$$\times \{(\alpha - n + 2) \ldots (\alpha - n + 2s)(\alpha + n + 2) \ldots (\alpha + n + 2s)\}^{-1}$$

corresponding to the choice $c_0 = \alpha + n$.

We can rewrite w in the form

$$w = z^\alpha \sum_{s=0}^{n-1} (-1)^s(\alpha + n)z^{2s}/\{(\alpha - n + 2) \ldots (\alpha - n + 2s)$$
$$\times (\alpha + n + 2) \ldots (\alpha + n + 2s)\} + (-1)^n z^{\alpha+2n}$$
$$\times \{(\alpha - n + 2) \ldots (\alpha + n - 2)(\alpha + n + 2) \ldots (\alpha + 3n)\}^{-1}$$
$$\times \left[1 + \sum_{s=1}^{\infty} (-1)^s z^{2s}/\{(\alpha + n + 2) \ldots (\alpha + n + 2s)\right.$$
$$\left.\times (\alpha + 3n + 2) \ldots (\alpha + 3n + 2s)\}\right].$$

We verify that $(w)_{\alpha=-n} = -J_n(z)/2^{2n-1}(n - 1)!$ and

$$\left(\frac{\partial w}{\partial \alpha}\right)_{\alpha=-n} = (\log z)\,(w)_{\alpha=-n}$$
$$+ z^{-n} \sum_{s=0}^{n-1} (-1)^s z^{2s}/\{(2 - 2n) \ldots (2s - 2n).2 \ldots (2s)\}$$
$$-(-1)^n z^n \phi(z)/\{(2 - 2n) \ldots (-2).2 \ldots (2n)\},$$

where

$$\phi(z) = \frac{1}{2n} + \sum_{s=1}^{\infty}(-1)^s z^{2s}\left[\frac{1}{2n} + \sum_{m=1}^{s}\left(\frac{1}{2m} + \frac{1}{2n+2m}\right)\right]$$
$$\times \{2\ldots(2s)(2n+2)\ldots(2n+2s)\}^{-1}$$

Accordingly,

$$-2^{2n-1}(n-1)!\left(\frac{\partial w}{\partial \alpha}\right)_{\alpha=-n} = J_n \log z - \tfrac{1}{2}\sum_{s=0}^{n-1}\frac{(n-s-1)!}{s!}\left(\frac{z}{2}\right)^{-n+2s}$$
$$-\tfrac{1}{2}\sum_{s=0}^{\infty}\frac{(-1)^s}{s!\,(n+s)!}\left(\frac{z}{2}\right)^{n+2s}\left\{\sum_{r=1}^{s}\frac{1}{r} + \sum_{r=0}^{s}\frac{1}{n+r}\right\}.$$

On subtraction of $\frac{1}{2}\sum_{r=1}^{n-1}(1/r)J_n(z)$ we obtain Bessel's function of the second kind, also called the Neumann function,

$$Y^{(n)}(z) = J_n(z)\log z - \tfrac{1}{2}\sum_{s=0}^{n-1}\frac{(n-s-1)!}{s!}\left(\frac{z}{2}\right)^{-n+2s}$$
$$-\tfrac{1}{2}\sum_{s=0}^{\infty}\frac{(-1)^s}{s!\,(n+s)!}\left(\frac{z}{2}\right)^{n+2s}\left\{\sum_{r=1}^{s}\frac{1}{r} + \sum_{r=1}^{n+s}\frac{1}{r}\right\}. \tag{1.6.2}$$

ILLUSTRATION 8

The differential equation

$$zw'' + (2-z)w' - w = 0$$

has exponents 0 and -1 at the origin, and the recurrence relationship is

$$(s+\alpha)(s+\alpha+1)c_s = (s+\alpha)c_{s-1}, \qquad (s \geqslant 1).$$

Corresponding to the larger exponent $\alpha = 0$ we have $c_s = c_{s-1}/(s+1)$ since the factor $s \neq 0$ for $s \geqslant 1$. This leads with $c_0 = 1$ to the solution

$$1 + \sum_{1}^{\infty} z^s/(s+1)! = (e^z - 1)/z.$$

On the other hand, the smaller exponent $\alpha = -1$ yields $c_s = c_{s-1}/s$ except when $s = 1$, since in this case the recurrence relationship is satisfied identically. Consequently c_1 is indeterminate and we have the solution

$$z^{-1}\left(c_0 + c_1\sum_{1}^{\infty} z^s/s!\right)$$
$$= \frac{c_0}{z} + \frac{c_1(e^z-1)}{z}.$$

Thus we have obtained two independent solutions $1/z$ and $(e^z - 1)/z$ both regular for $|z| > 0$. Despite the fact that the exponents differ by an integer, there is no logarithmic term in the solution.

1.7 Singularities at infinity

To discuss the behaviour of the solution of the differential equation near infinity we make the transformation $z = 1/Z$ and so reduce the problem to the behaviour of the transformed equation near the origin. A simple calculation shows that

$$w'' + p(z)w' + q(z)w = 0 \tag{1.7.1}$$

transforms into

$$Z^4\frac{d^2w}{dZ^2} + \left\{2Z^3 - Z^2p\left(\frac{1}{Z}\right)\right\}\frac{dw}{dZ} + q\left(\frac{1}{Z}\right)w = 0. \tag{1.7.2}$$

We define the point at infinity to be an ordinary point, a regular singularity or an irregular singularity of (1.7.1), according as the origin is an ordinary point, a regular singularity or an irregular singularity of (1.7.2).

It follows that (i) the point at infinity is an ordinary point of (1.7.1) if $\{p(1/Z) - 2Z\}/Z^2$ and $q(1/Z)/Z^4$ are regular in the neighbourhood of $Z = 0$, (ii) the point at infinity is a regular singularity if $p(1/Z)/Z$ and $q(1/Z)/Z^2$ are regular in the neighbourhood of the origin.

At ordinary and regular points we may apply the preceding theory and so obtain regular expansions in powers of $1/z$.

ILLUSTRATION 9

The transformation $z = 1/Z$ changes Legendre's differential equation of illustration 1 into

$$Z^2(Z^2 - 1)\frac{d^2w}{dZ^2} + 2Z^3\frac{dw}{dZ} + \nu(\nu + 1)w = 0$$

and so the point at infinity is a regular singularity.

The substitution $w = Z^\alpha \sum c_s Z^s$ yields

$$\sum_{s=0}^{\infty} c_s(s + \alpha + \nu)(s + \alpha - \nu - 1)Z^{s+\alpha} = \sum_{s=0}^{\infty} c_s(s + \alpha)(s + \alpha + 1)Z^{s+\alpha+2}$$

and so

$$(\alpha + \nu)(\alpha - \nu - 1)c_0 = 0,$$
$$(\alpha + \nu + 1)(\alpha - \nu)c_1 = 0,$$
$$(\alpha + \nu + s)(\alpha - \nu - 1 + s)c_s = (\alpha + s - 2)(\alpha + s - 1)c_{s-2}, \quad (s \geqslant 2).$$

The exponents at infinity are $-\nu$ and $\nu + 1$, and the second equation gives $c_1 = 0$ except in the case $\nu = 0$, when c_1 is indeterminate.

When $2\nu + 1$ is *not* an integer the recurrence relations corresponding to the exponents $-\nu$ and $\nu + 1$ are

$$s(s - 2\nu - 1)c_s = (s - \nu - 1)(s - \nu - 2)c_{s-2} \qquad (1.7.3)$$

and

$$s(s + 2\nu + 1)c_s = (s + \nu - 1)(s + \nu)c_{s-2}$$

and so the solutions of Legendre's differential equation near infinity are

$$w_1(z) = z^{\nu} \sum_{s=0}^{\infty} \frac{(-1)^s \, \Gamma(\nu + 1) \, \Gamma(\nu + \frac{1}{2} - s)}{s! \, \Gamma(\nu + \frac{1}{2}) \, \Gamma(\nu - 2s + 1)} \frac{1}{(2z)^{2s}} \qquad (1.7.4)$$

and

$$w_2(z) = z^{-\nu-1} \sum_{s=0}^{\infty} \frac{\Gamma(\nu + 2s + 1) \, \Gamma(\nu + \frac{3}{2})}{s! \, \Gamma(\nu + 1) \, \Gamma(\nu + s + \frac{3}{2})} \frac{1}{(2z)^{2s}}. \qquad (1.7.5)$$

These solutions $w_1(z)$ and $w_2(z)$ both exist even when ν is an integer and correspond to (1.6.1) with $c = 0$.

However, if ν is half an odd positive integer, then the presence of $\Gamma(\nu + \frac{1}{2} - s)$ in the numerator of $w_1(z)$ shows that this solution must be replaced by a logarithmic solution according to the method of section 1.6. If ν is half an odd negative integer, the solution $w_1(z)$ exists, but $w_2(z)$ must be replaced by a logarithmic solution.

With these exceptions, we see that $z^{-\nu}w_1(z)$ and $z^{\nu+1}w_2(z)$ are regular for $|Z| < 1$; that is, for $|z| > 1$.

We see from the recurrence relation (1.7.3) that $c_s = 0$ when $s = \nu + 1$. It follows that if ν is an integer n the solution $w_1(z)$ is a polynomial. On the choice of the multiplier $(2n)!/\{2^n(n!)^2\}$ we retrieve the Legendre polynomial introduced in illustration 1.

Exercises

Obtain series solutions in ascending powers of z of the following differential equations.

1. $w'' - zw = 0$.
2. $(1 - z^2)w'' - zw' + n^2w = 0$ (Chebyshev). In this exercise show that polynomial solutions exist.
3. $2zw'' + (3 - 2z)w' - 2w = 0$.
4. $zw'' - w' + 4z^3w = 0$.
5. $z^2w'' + (z - 1)(zw' - w) = 0$.
6. $zw'' + w' + zw = 0$.
7. $zw'' + zw' + w = 0$.
8. $zw'' + (p - z)w' - w = 0$ in the cases (a) $p = 1$, (b) $p = 2$.

Examine the nature of the point at infinity with respect to the following differential equations.

9. $z^3(z-1)w'' + 2z(z^2-2)w' + 3w = 0$.
10. $2z^3w'' + (z+z^2)w' + w = 0$.
11. $z^2w'' + zw' + (z^2-\nu^2)w = 0$ (Bessel).
12. $(1-z^2)w'' - zw' + n^2w = 0$ (Chebyshev).
13. $w'' - 2zw' + 2\nu w = 0$ (Hermite).
14. $zw'' + (\alpha+1-z)w' + \nu w = 0$ (Laguerre).
15. $z(1-z)w'' + \{c-(a+b+1)z\}w' - abw = 0$ (hypergeometric).

ANSWERS

1. $A\left[1+\sum_{s=1}^{\infty} z^{3s}/\{2.3.5.6...(3s-1)(3s)\}\right] + B\left[z+\sum_{s=1}^{\infty} z^{3s+1}/\{3.4.6.7...(3s)(3s+1)\}\right]$.
2. $A[1-n^2z^2/2!-(2^2-n^2)n^2z^4/4!-(4^2-n^2)(2^2-n^2)n^2z^6/6!\ldots]$ $+B[z+(1-n^2)z^3/3!+(3^2-n^2)(1-n^2)z^5/5!+\ldots]$.
3. $Az^{-1/2}e^z + B\sum_{s=0}^{\infty} 2^{2s}s!z^s(2s+1)!$.
4. $A\cos(z^2) + B\sin(z^2)$.
5. $Az + B\left[z\log z + \sum_{s=1}^{\infty}(-1)^{s-1}z^s/[(s-1)\{(s-1)!\}]\right]$.
6. $(A+B\log z)\sum_{s=0}^{\infty}(-1)^s z^{2s}/\{2^{2s}(s!)^2\} + B\sum_{s=0}^{\infty}\left(\sum_{m=0}^{s}\frac{1}{m}\right)(-1)^{s-1}z^{2s} \times \{2^{2s}(s!)^2\}^{-1}$.
7. $(A+B\log z)ze^{-z} - B\left\{1+\sum_{s=2}^{\infty}\left(\sum_{m=1}^{s-1}\frac{1}{m}\right)(-1)^{s-1}z^s/(s-1)!\right\}$.
8. (a) $(A+B\log z)e^z - B\sum_{s=1}^{\infty}\left(\sum_{m=1}^{s}\frac{1}{m}\right)z^s/s!$.

   (b) $z^{-1}(Ae^z+B)$.
9. Ordinary.

10, 12, 15. Regular.

11, 13, 14. Irregular.

CHAPTER 2

Contour Integral Solutions of an Ordinary Linear Differential Equation

2.1 General comments

Of the two principal methods of obtaining explicit solutions of ordinary linear differential equations, expanding the solution as a power series around a given point, and expressing the solution as a contour integral, the latter will in general convey the greater amount of information. Each method is only really useful in limited circumstances. The first method gives immediately a complete description of the solution at the given point, but to extract information about the solution at other points becomes progressively more difficult as the limit of convergence is approached. The information conveyed by a contour-integral solution requires greater interpretation, but by suitable distortions of the contour of integration it is frequently possible to isolate the particular form of data sought. It is also possible to derive expansions of the solution near irregular singular points, where the ordinary series-expansion methods cannot be applied.

In this chapter we shall confine our attention to a limited range of types of contour integral, which are therefore only applicable to a correspondingly limited range of differential equations, which include those of most interest in mathematical physics.

2.2 Laplace-type integrals

We restrict ourselves to equations whose coefficients are polynomials in the independent variable, and we look for solutions of the form

$$w(z) = \int_C e^{zt} f(t)\, dt, \tag{2.2.1}$$

where $f(t)$ is a function to be determined, and C is a contour of integration which at this stage is undefined.

We assume for the moment that the contour and the function $f(t)$ are so chosen that the integral is uniformly convergent and so we may differentiate under the integral sign. We may verify subsequently that the condition of uniform convergence is satisfied. Thus

$$w^{(k)}(z) = \int_C e^{zt} t^k f(t)\, dt, \tag{2.2.2}$$

where we write $w^{(k)}(z)$ for $d^k w/dz^k$.

All the equations to which analysis will be applied in the later chapters will be of second order, but as it is no more difficult to describe the theory for the equation of order n we consider the equation

$$\sum_{k=0}^{n} a_k(z) w^{(n-k)}(z) = 0, \tag{2.2.3}$$

where the $a_k(z)$ are polynomials in z, whose greatest degree is m, say. Hence if we substitute (2.2.2) into (2.2.3) we obtain

$$\int_C e^{zt} F(z, t) f(t)\, dt = 0 \tag{2.2.4}$$

where $F(z, t)$ is of degree m in z and of degree n in t. This can clearly be rearranged in the form

$$\sum_{r=0}^{m} z^r \int_C e^{zt} p_r(t) f(t)\, dt = 0. \tag{2.2.5}$$

Consider first the term

$$z \int_C e^{zt} p_1(t) f(t)\, dt.$$

If this is integrated by parts it becomes

$$\Delta_C e^{zt} p_1(t) f(t) - \int_C e^{zt} \frac{d}{dt}\{p_1(t) f(t)\}\, dt,$$

where Δ_C represents the change in value of the function in going from one end of the contour to the other. Similarly we obtain

$$\begin{aligned} z^2 \int_C e^{zt} p_2(t) f(t)\, dt &= z\Delta_C\, e^{zt} p_2(t) f(t)\, dt - z \int_C e^{zt} \frac{d}{dt}\{p_2(t) f(t)\}\, dt \\ &= \Delta_C\, e^{zt}\left[z p_2(t) f(t) - \frac{d}{dt}\{p_2(t) f(t)\}\right] + \int_C e^{zt} \frac{d^2}{dt^2}\{p_2(t) f(t)\}\, dt. \end{aligned}$$

Generally we see that

$$\begin{aligned} z^r \int_C e^{zt} p_r(t) f(t)\, dt = \Delta_C\, e^{zt}&\left[\sum_{s=0}^{r-1} (-1)^s z^{r-s-1} \frac{d^s}{dt^s}\{p_r(t) f(t)\}\right] \\ &+ (-1)^r \int_C e^{zt} \frac{d^r}{dt^r}\{p_r(t) f(t)\}\, dt. \end{aligned}$$

If these expansions are substituted in (2.2.5) we obtain

$$\Delta_C\, \mathrm{e}^{zt}\Phi(z, t) + \int_C \mathrm{e}^{zt}\left[\sum_{r=0}^{m}(-1)^r\frac{\mathrm{d}^r}{\mathrm{d}t^r}\{p_r(t)f(t)\}\right]\mathrm{d}t = 0 \qquad (2.2.6)$$

where

$$\Phi(z, t) = \sum_{r=1}^{m}\sum_{s=0}^{r-1}(-1)^s z^{r-s-1}\frac{\mathrm{d}^s}{\mathrm{d}t^s}\{p_r(t)f(t)\}.$$

We select the function $f(t)$ to satisfy the differential equation

$$\sum_{r=0}^{m}(-)^r\frac{\mathrm{d}^r}{\mathrm{d}t^r}\{p_r(t)f(t)\} = 0, \qquad (2.2.7)$$

and then having fixed $f(t)$ we select a contour C such that

$$\Delta_C\, \mathrm{e}^{zt}\Phi(z, t) = 0. \qquad (2.2.8)$$

Certain comments are in order. We have in essence replaced the problem of solving (2.2.3), a differential equation of order n, whose coefficients are polynomials of degree m, by an equivalent problem of solving (2.2.7), a differential equation of order m whose coefficients are polynomials of degree n. In general it is more desirable to reduce the order of an equation than to reduce the degree of its coefficients. It follows that this technique is unlikely to be useful if $m \geqslant n$. Of course it might happen that the equation of higher order had an obvious solution, while that of lower order had not, but this situation is highly unusual.

If the chosen contour is one that can be distorted into a line parallel with the imaginary axis to the right of all the singularities of the integrand, then (2.2.1) can be regarded as an inversion-type integral of a Laplace transform. This contour is often associated with the name of Bromwich. It follows that any solution of the equation which can be obtained by using the Laplace transform can be obtained from the above analysis. On the other hand, other solutions may be obtained by our analysis which gives contours that cannot be distorted into the Bromwich contour, and these would be unobtainable from the application of a Laplace transform. Thus this particular type of contour integral can be regarded as a generalization of the Laplace transform.

2.3 The differential equation with constant coefficients

Though the solution of such an equation can be obtained by more elementary methods, it is instructive to follow the analysis. In this case, if the equation is

$$\sum_{k=0}^{n} a_k\frac{\mathrm{d}^{n-k}w}{\mathrm{d}z^{n-k}} = 0, \qquad (2.3.1)$$

then

$$p_0(t) = \sum_{k=0}^{n} a_k t^{n-k}, \tag{2.3.2}$$

and

$$p_r(t) \equiv 0, \qquad r \geqslant 1. \tag{2.3.3}$$

Thus

$$\Phi(z, t) \equiv 0, \tag{2.3.4}$$

and hence any contour giving a convergent integral will suffice. It is clear in fact that

$$\int_C e^{zt} p_0(t) f(t)\, dt = 0$$

provided C lies entirely within a part of the plane in which $p_0(t)f(t)$ is regular. Suppose $p_0(t)$ has a factor $(t - \alpha)^j$. Then we can allow $f(t)$ to have a pole of order j at $t = \alpha$, and $p_0(t)f(t)$ will still be regular near $t = \alpha$. Then near $t = \alpha$

$$f(t) = \beta_j(t - \alpha)^{-j} + \beta_{j-1}(t - \alpha)^{-j+1} + \ldots + \beta_1(t - \alpha)^{-1} + \phi(t)$$

where $\phi(t)$ is regular at $t = \alpha$. We can select as a contour a small circle surrounding $t = \alpha$, and then a solution is

$$\begin{aligned} w(z) &= \int_C e^{zt} f(t)\, dt = e^{\alpha z} \sum_{i=0}^{j-1} \beta_{j-i} \int_C (t - \alpha)^{i-j}\, e^{z(t-\alpha)}\, dt \\ &= e^{\alpha z} \sum_{i=0}^{j-1} \beta_{j-i} \frac{z^{j-i-1}}{\{(j-i)!\}}. \end{aligned} \tag{2.3.5}$$

Treating all the factors of $p_0(t)$ in this way, we get a general solution of the equation.

2.4 The Laplace linear differential equation

The next differential equation with polynomial coefficients in order of simplicity is the Laplace linear differential equation

$$\sum_{k=0}^{n} (a_k + b_k z) \frac{d^{n-k} w}{dz^{n-k}} = 0, \tag{2.4.1}$$

where the coefficients are all linear functions of z.

In this case the only pair of $p_r(t)$ different from zero are respectively

$$p_0(t) = \sum_{k=0}^{n} a_k t^{n-k},$$

and

$$p_1(t) = \sum_{k=0}^{n} b_k t^{n-k}.$$

The equation (2.2.7) reduces to

$$\frac{\mathrm{d}}{\mathrm{d}t}\{p_1(t)f(t)\} - p_0(t)f(t) = 0 \tag{2.4.2}$$

and hence if we write

$$g(t) = p_1(t)f(t),$$

then

$$\frac{\mathrm{d}}{\mathrm{d}t}\log g(t) = p_0(t)/p_1(t). \tag{2.4.3}$$

Thus

$$g(t) = \exp\left[\int_{t_0}^{t}\{p_0(\tau)/p_1(\tau)\}\,\mathrm{d}\tau\right]. \tag{2.4.4}$$

The corresponding condition on the contour reduces to

$$\Delta_C\, \mathrm{e}^{zt}g(t) = 0. \tag{2.4.5}$$

We can ensure this by choosing for C a closed contour round which $g(t)$ is a single-valued function; that is, it returns to its starting value after traversing the contour once. We may equally achieve this by ensuring that $\mathrm{e}^{zt}g(t)$ vanishes at both ends of the contour. Thus we may choose as an end point of the contour a finite point if it is a zero of $g(t)$, or we can choose infinity as an end point, provided we select a direction of approach to infinity for the contour along which $\mathrm{e}^{zt}g(t)$ tends uniformly to zero.

Clearly we may assume that a_0 and b_0 are not both zero. Suppose first $a_0 = 0$, so that the degree of $p_0(t)$ is less than that of $p_1(t)$. Then we can resolve their quotient uniquely into partial fractions of the form

$$\frac{\beta_j}{(t-\alpha)^j} + \frac{\beta_{j-1}}{(t-\alpha)^{j-1}} + \ldots + \frac{\beta_1}{(t-\alpha)}$$

and $g(t)$ will have a factor

$$(t-\alpha)^{\beta_1}\exp\left[-\left\{\frac{\beta_j}{j-1}(t-\alpha)^{1-j} + \frac{\beta_{j-1}}{j-2}(t-\alpha)^{2-j} + \ldots \right.\right.$$
$$\left.\left. + \beta_2(t-\alpha)^{-1}\right\}\right]. \tag{2.4.6}$$

If $j = 1$, the exponential factor does not appear and we can choose as a contour one that starts from infinity, goes once in the positive direction round $t = \alpha$, and returns to infinity. Since the dominant factor of $\mathrm{e}^{zt}g(t)$ near infinity will be e^{zt}, we must choose a direction of approach to infinity so that $\mathscr{R}(zt) < 0$. If β_1 is a negative integer, the singularity at $t = \alpha$ is a pole and we may reduce the contour to a closed curve round $t = \alpha$. If $\beta_1 > 0$, $t = \alpha$ is a zero of $g(t)$ and the contour above can be replaced by one from infinity to $t = \alpha$.

If $j > 1$, then near $t = \alpha$ the dominant factor of $g(t)$ will be

$$\exp\left\{-\frac{\beta_j}{j-1}(t-\alpha)^{1-j}\right\}$$

and hence there are $j - 1$ sectors of approach to $t = \alpha$ on which the exponential will tend to zero as we approach $t = \alpha$, and $j - 1$ sectors in which the exponential tends to infinity. In this case, in addition to the above contour from infinity round $t = \alpha$ and back to infinity, we can choose another $j - 1$, starting from infinity and terminating at $t = \alpha$, in one of the $j - 1$ sectors in which the exponential vanishes.

In any case, associated with a factor $(t - \alpha)^j$ of $p_1(t)$ there are j suitable contours, and hence we can find n independent solutions of the differential equation.

If neither a_0 nor b_0 is zero, then $p_0(t)$ and $p_1(t)$ are of the same degree, and so

$$p_0(t) = \frac{a_0}{b_0}p_1(t) + p_0^*(t)$$

where $p_0^*(t)$ is of smaller degree than $p_1(t)$. Hence this case reduces essentially to the previous one, wherein z is replaced by $z + a_0/b_0$. It is clear therefore that once again we can obtain n independent solutions.

If $b_0 = 0$, and the degree N of $p_1(t)$ is actually less than the degree n of $p_0(t)$, we have

$$p_0(t) = p_1(t)\left(\sum_{i=0}^{n-N} c_i t^i\right) + \hat{p}_0(t)$$

where the degree of $\hat{p}_0(t)$ is less than that of $p_1(t)$, and

$$g(t) = \exp\left[\sum_{i=0}^{n-N}\frac{c_i}{i+1}t^{i+1} + \int_{t_0}^{t}\{\hat{p}_0(\tau)/p_1(\tau)\}\,d\tau\right]. \qquad (2.4.7)$$

The last part can clearly be dealt with in the same way as before. Now the dominant part of $e^{zt}g(t)$ at infinity will be

$$\exp\left\{\frac{c_{n-N}}{n-N+1}t^{n-N+1}\right\}.$$

The real part of $c_{n-N}t^{n-N+1}$ will tend to plus infinity in $n - N + 1$ sectors as $t \to \infty$, and will tend to minus infinity in the complementary $n - N + 1$ sectors. The direction of approach to infinity must be chosen in one of the latter sectors, and this enables us to select n independent solutions.

We see therefore that it is always possible by using a contour integral of the Laplace type to obtain the general solution of a Laplace linear differential equation.

2.5 Illustrations of the method

ILLUSTRATION 1

The confluent hypergeometric equation is

$$zw'' + (c - z)w' - aw = 0. \tag{2.5.1}$$

Then

$$p_0(t) = ct - a \quad \text{and} \quad p_1(t) = t(t - 1)$$

and so

$$\frac{p_0(t)}{p_1(t)} = \frac{c - a}{t - 1} + \frac{a}{t}.$$

Hence

$$g(t) = t^a(t - 1)^{c-a}$$

and

$$w = \int_C \mathrm{e}^{zt} t^{a-1}(t - 1)^{c-a-1}\, \mathrm{d}t \tag{2.5.2}$$

is a solution, provided

$$\Delta_C\, \mathrm{e}^{zt} t^a(t - 1)^{c-a} = 0. \tag{2.5.3}$$

If a is zero or a negative integer there is a pole at $t = 0$, and a small circle round $t = 0$ is a possible contour (Fig. 2.1(a)). Similarly if $c - a$ is zero or a negative integer, a small circle round the point $t = 1$ will suffice. If c is an integer, we can choose a simple closed curve which has both $t = 0$ and $t = 1$ in its interior (Fig. 2.1(b)). As a single circuit of the curve would increase the argument of $(t - 1)^{c-a}$ by $2(c - a)\pi$ and increase arg t^a by $2\pi a$, it would increase arg $g(t)$ by $2\pi c$ and so $g(t)$ would be single-valued on C. If both $\mathscr{R}(a) > 0$ and $\mathscr{R}(c - a) > 0$, $g(t)$ has zeros at 0 and 1, and we can choose a line segment joining these points (Fig. 2.1.(c)). In the general case we could choose a closed contour, as we shall show later in this chapter, but it is possibly easier to choose one of the infinite contours shown in Fig. 2.1.

Any other solution will require a contour extending to infinity. In particular, if $\mathscr{R}(z) > 0$, we can choose two contours, each starting and finishing as asymptotes to the negative real t-axis, one round the point $t = 1$ and the other round $t = 0$. If $\mathscr{R}(z) < 0$ we would correspondingly choose a contour asymptotic to the positive real axis (Fig. 2.1(d)). If $\mathscr{R}(c - a) > 0$ we could terminate the contour at $t = 1$ without returning it to infinity, and if $\mathscr{R}(a) > 0$ we could terminate it at $t = 0$. This will give two independent solutions in all cases.

The particular case $c = \alpha$, $a = -\nu$ gives the Laguerre equation.

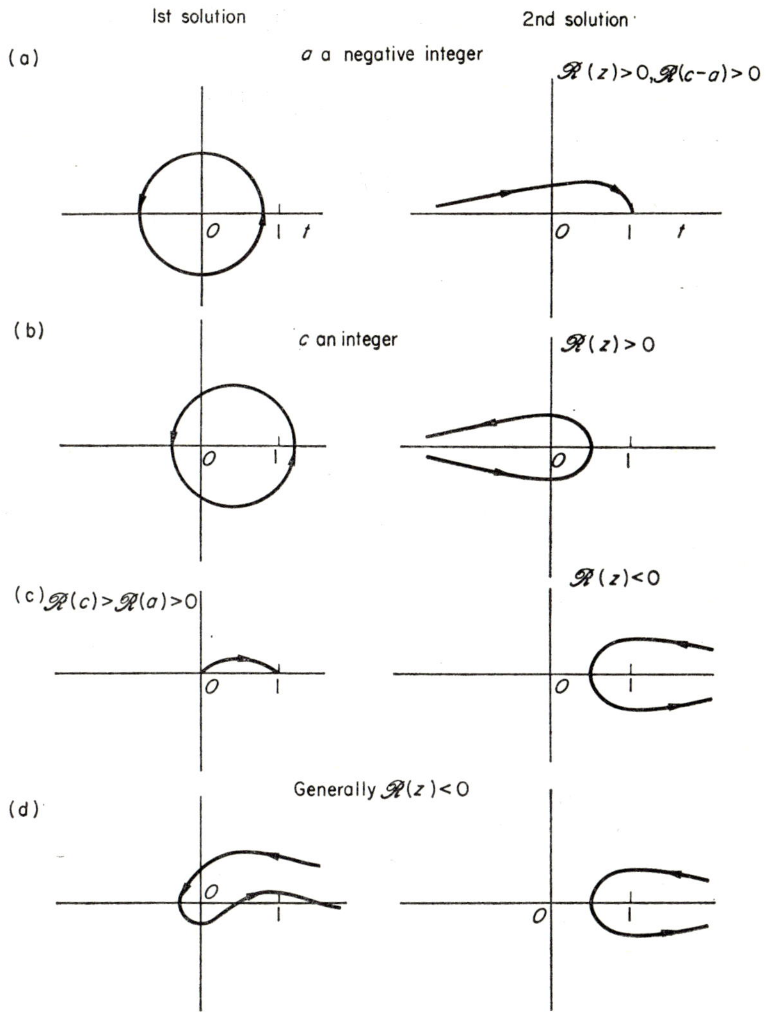

Fig. 2.1.

ILLUSTRATION 2

Consider the equation

$$zw'' + aw' + b^2zw = 0, \tag{2.5.4}$$

where we shall assume that a and b are real.

Then

$$p_0(t) = at \quad \text{and} \quad p_1(t) = t^2 + b^2$$

and so

$$g(t) = (t^2 + b^2)^{a/2}.$$

Hence

$$w(z) = \int_C e^{zt}(t^2 + b^2)^{a/2-1}\, dt \tag{2.5.5}$$

is a solution, provided

$$\Delta_C\, e^{zt}(t^2 + b^2)^{a/2} = 0. \tag{2.5.6}$$

If a is zero or a negative even integer, there are poles of the integrand at $\pm ib$. Hence two independent solutions are given by closed curves round $t = \pm ib$ (Fig. 2.2(a)). If a is a positive even integer, the integrand

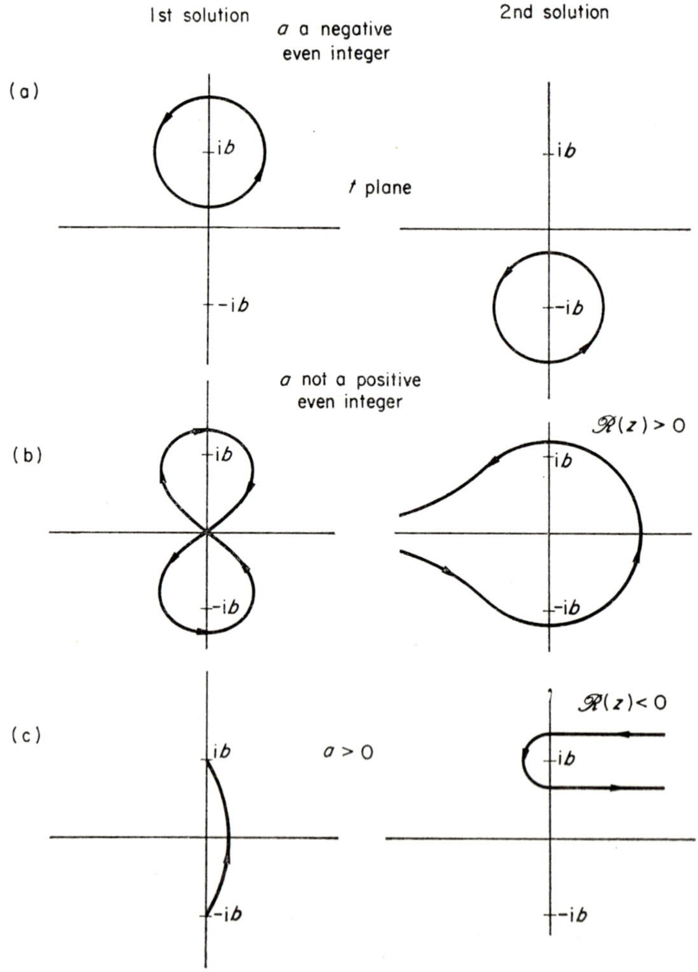

Fig. 2.2.

is an entire function of t, and thus any finite contour will give the trivial solution, $w = 0$. For any other values of a, a single loop round $t = ib$ excluding $t = -ib$ will increase $\arg g(t)$ by $a\pi$, while a single loop round $t = -ib$ excluding $t = ib$ will also increase $\arg g(t)$ by $a\pi$. Hence a figure of eight going once round $t = ib$ in one direction and round $t = -ib$ in the other restores $g(t)$ to its initial value, and so is a suitable curve (see Fig. 2.2(b)). If $a > 0$ this can be contracted to the line joining $\pm ib$ (Fig. 2.2(c)).

All other solutions must start or finish at infinity.

In the particular case $b = 1$, $a = 2\nu + 1$, we have an equation closely allied to the Bessel equation.

Illustration 3

Consider the equation

$$zw'' - (2z - \alpha)w' + (z + \beta)w = 0, \tag{2.5.7}$$

where again we shall assume that α and β are real, and $\alpha + \beta \neq 0$. Then

$$p_0(t) = \alpha t + \beta \quad \text{and} \quad p_1(t) = (t - 1)^2$$

and so

$$g(t) = (t - 1)^{\alpha} \exp\left\{-\frac{\alpha + \beta}{t - 1}\right\}.$$

Thus

$$w(z) = \int_C (t - 1)^{\alpha - 2} \exp\left\{zt - \frac{\alpha + \beta}{t - 1}\right\} \mathrm{d}t \tag{2.5.8}$$

is a solution, provided

$$\Delta_C (t - 1)^{\alpha} \exp\left\{zt - \frac{\alpha + \beta}{t - 1}\right\} = 0. \tag{2.5.9}$$

If α is an integer, a closed curve round $t = 1$ (Fig. 2.3(a)) gives an acceptable solution, and if not, this is generalized to a solution starting from infinity in a direction for which $\mathcal{R}(zt) < 0$, going once round $t = 1$

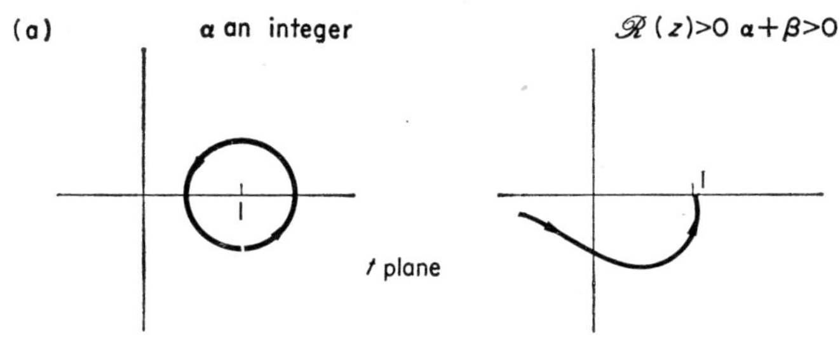

Fig. 2.3(a).

and returning to infinity in the same direction (Fig. 2.3(b)). If $\alpha + \beta > 0$ we can choose a second solution starting from infinity in this same direction, and terminating at $t = 1$, approaching $t = 1$ along the positive

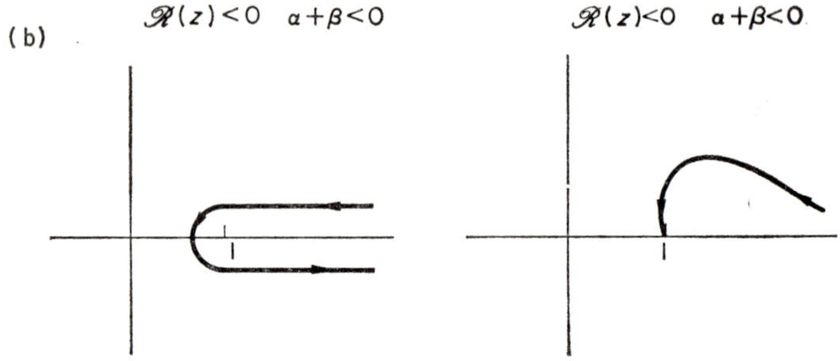

Fig. 2.3(b)

real axis. Similarly if $\alpha + \beta < 0$ the approach would be along the negative real axis. In either case the second solution is independent of the first.

ILLUSTRATION 4

The Hermite equation is

$$w'' - 2zw' + 2\nu w = 0. \tag{2.5.10}$$

In this case

$$p_0(t) = t^2 + 2\nu \quad \text{and} \quad p_1(t) = -2t$$

and so

$$g(t) = t^{-\nu} \exp(-t^2/4).$$

Thus

$$w(z) = \int_C t^{-\nu-1} \exp(zt - t^2/4) \tag{2.5.11}$$

is a solution, provided

$$\Delta_C t^{-\nu} \exp(zt - t^2/4) = 0. \tag{2.5.12}$$

If ν is zero or a positive integer, the integrand has a pole at $t = 0$, and an acceptable contour is a simple closed curve round the origin (Fig. 2.4(a)). If ν is a negative integer, the integrand is an entire function of t, and any finite closed contour must give the trivial solution $w = 0$. If ν is not an integer, the only finite singularity is a branch at point $t = 0$, and so every other contour giving a non-trivial solution must be infinite.

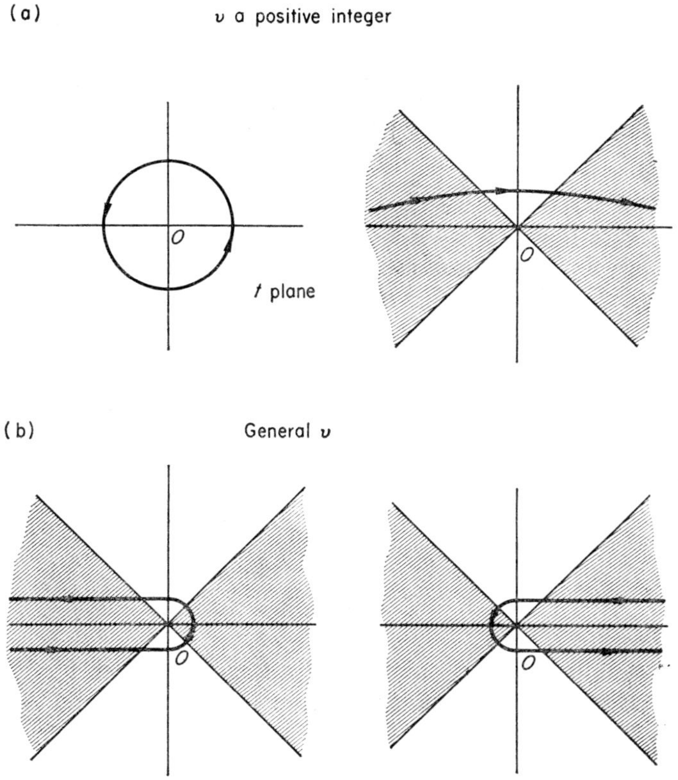

Fig. 2.4.

If $\nu < 0$ they can terminate at $t = 0$, but if $\nu \geqslant 0$ they must start and finish at infinity. At infinity the dominant term is now

$$\exp(-t^2/4)$$

and so we require a direction for the approach to infinity on which $\mathscr{R}(t^2) > 0$, that is, $-\pi/4 < \arg t < \pi/4$ or $3\pi/4 < \arg t < 5\pi/4$. Thus two independent solutions are given by contours, one starting from infinity on the negative real axis, going once round $t = 0$ and returning to its starting point, and the other starting from infinity on the positive real axis, encircling the origin and returning to its starting point.

ILLUSTRATION 5

The Airy equation is

$$w'' + zw = 0. \tag{2.5.13}$$

In this final example

$$p_0(t) = t^2 \quad \text{and} \quad p_1(t) = 1$$

and hence

$$g(t) = \tfrac{1}{3}t^3.$$

Thus a solution is

$$w(z) = \int_C \exp\,(zt + \tfrac{1}{3}t^3)\,\mathrm{d}t, \tag{2.5.14}$$

provided

$$\Delta_C \exp\,(zt + \tfrac{1}{3}t^3) = 0. \tag{2.5.15}$$

In this case the integrand is always an entire function of t. Consequently only infinite contours can be chosen, and since there are no finite zeros, the contours must both begin and end at infinity. At infinity the dominant term is

$$\exp\,(\tfrac{1}{3}t^3)$$

and hence we must choose a direction of approach to infinity on which

$$\mathscr{R}(t^3) < 0,$$

that is,

$$(2k + \tfrac{1}{2})\pi < \arg t^3 < (2k + \tfrac{3}{2})\pi$$

or

$$\frac{4k+1}{6}\pi < \arg t < \frac{4k+3}{6}\pi$$

for, say, $k = -1, 0, 1$.

Three contours of integration suggest themselves (see Fig. 2.5); C_1, C_2 and C_3. Only two independent solutions are possible, and there must

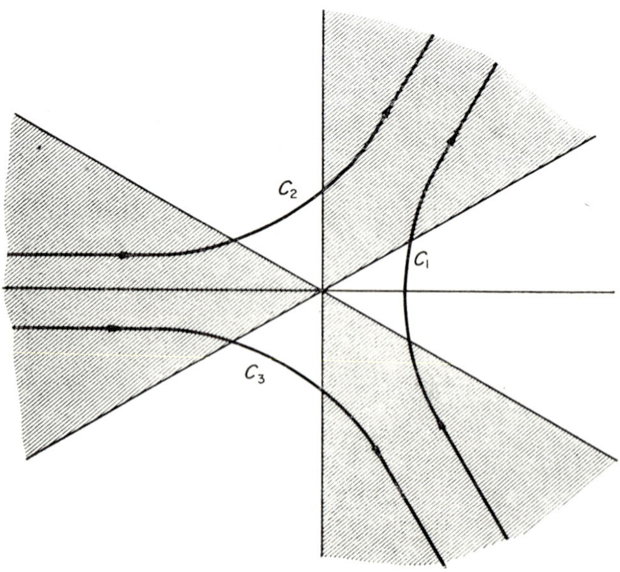

Fig. 2.5.

be a relation between the solutions. If we call $w_i(z)$ the solution associated with C_i, then it is clear from the contours that

$$w_2(z) = w_3(z) + w_1(z). \tag{2.5.16}$$

The solutions conventionally taken are

$$\frac{1}{2\pi i} w_1(z) \quad \text{and} \quad \tfrac{1}{2}\{w_2(z) + w_3(z)\}.$$

In fact we may distort C_1 continuously into the imaginary axis, and so

$$w_1(z) = 2i \int_0^\infty \cos (zu - u^3/3)\, du. \tag{2.5.17}$$

We may similarly distort C_2 and C_3 into the negative real axis and half the imaginary axis so that

$$w_2(z) + w_3(z) = 2 \int_0^\infty \sin (zu - u^3/3)\, du$$
$$- 2 \int_0^\infty \exp \{-(zu + u^3/3)\}\, du. \tag{2.5.18}$$

2.6 Generalization of the Laplace-type integral

An obvious generalization is to replace the integral (2.2.1) by

$$w(z) = \int_C e^{th(z)} f(t)\, dt \tag{2.6.1}$$

where $h(z)$ is some given function of z, preferably an entire function. It is clear that as the complexity of the function $h(z)$ is increased, the class of differential equation for which its application would be suitable will decrease. We might reasonably anticipate that $h(z) = z^2$, which is the next elementary case, provides the greatest interest. Let us apply this to two of the illustrations of the last section.

Consider the equation

$$zw'' + aw' + b^2 zw = 0. \tag{2.6.2}$$

Now if

$$w = \int_C \exp (tz^2) f(t)\, dt,$$

then

$$w' = 2z \int_C t \exp (tz^2) f(t)\, dt,$$

and

$$w'' = \int_C \exp (tz^2)(4z^2t^2 + 2t) f(t)\, dt.$$

Hence

$$\int_C \exp(tz^2)(4z^2t^2 + 2t + 2at + b^2)f(t) = 0$$

and integrating the first term by parts gives us

$$4\Delta_C \exp(tz^2)t^2f(t) + \int_C \exp(tz^2)\left[\{2(1 + a)t + b^2\}f(t) - 4\frac{\mathrm{d}}{\mathrm{d}t}\{t^2f(t)\}\right] \mathrm{d}t = 0.$$

We proceed as before and put $g(t) = t^2f(t)$, and choose $g(t)$ to satisfy

$$4\frac{\mathrm{d}g}{\mathrm{d}t} - \{2(1 + a)t + b^2\}t^{-2}g = 0 \tag{2.6.3}$$

so that

$$g(t) = t^{\frac{1}{2}(1+a)} \exp(-b^2/4t). \tag{2.6.4}$$

Then the contour is selected so that

$$\Delta_C \exp(z^2t)g(t) = 0. \tag{2.6.5}$$

There are two obvious choices for a contour. The first, starting from infinity in a direction for which $\mathcal{R}(z^2t) < 0$, would go once round the origin and return to infinity. Since this contour can be kept a finite distance from the only singularity of the integrand which is at the origin, it must result in a solution which is an entire function of z and is clearly also an even function. The second contour, starting from infinity in the same direction, terminates at the origin, approaching it along the positive real axis.

As a second example take illustration 4 above, the Hermite equation

$$w'' - 2zw' + 2\nu w = 0, \tag{2.6.6}$$

which would give correspondingly

$$\int_C \exp(tz^2)\{4z^2(t - 1)t + 2(t + \nu)\}f(t)\,\mathrm{d}t = 0. \tag{2.6.7}$$

In this case

$$g(t) = (t - 1)tf(t)$$

so that

$$\frac{2}{g}\frac{\mathrm{d}g}{\mathrm{d}t} = \frac{1 + \nu}{t - 1} - \frac{\nu}{t}.$$

Then

$$w(z) = \int_C t^{-1-\nu/2}(t - 1)^{(\nu-1)/2} \exp(z^2t)\,\mathrm{d}t \tag{2.6.8}$$

is a solution, provided

$$\Delta_C \exp(z^2t)(t - 1)^{(\nu+1)/2}t^{-\nu/2} = 0.$$

The choice of contour in this case is essentially the same as that in illustration 1 above.

2.7 The Euler contour integral

The only other form of contour integral we wish to consider in this chapter is a generalization to the complex plane of a real integral, which in that form is usually called the Euler transformation.

We consider integrals of the form

$$w(z) = \int_C (t - z)^\mu f(t)\,\mathrm{d}t \tag{2.7.1}$$

where μ is a parameter to be determined, as are also the unknown function $f(t)$ and the contour C.

Provided the contour is chosen to ensure the uniform convergence of the integral, we may again differentiate under the integral sign to obtain

$$w^{(k)}(z) = (-1)^k \mu(\mu - 1) \ldots (\mu - k + 1) \int_C (t - z)^{\mu-k} f(t)\,\mathrm{d}t. \tag{2.7.2}$$

Let us suppose $q(z)$ is a polynomial of degree $n - r$, and denote the usual binomial coefficient by $\binom{\nu}{k}$. Then consider

$$\begin{aligned}\sum_{k=0}^{n-r} (-1)^k \binom{\mu + k + r - n}{k} q^{(k)}(z) w^{(n-r-k)}(z) \\ = (-1)^{n-r}\mu(\mu - 1) \ldots (\mu + r - n + 1) \\ \times \int_C (t - z)^{\mu+r-n} \left[\sum_{k=0}^{n-r} \frac{q^{(k)}(z)}{k!}(t - z)^k\right] f(t)\,\mathrm{d}t \\ = (-1)^{n-r}\mu(\mu - 1) \ldots (\mu + r - n + 1) \\ \times \int_C (t - z)^{\mu+r-n} q(t) f(t)\,\mathrm{d}t.\end{aligned} \tag{2.7.3}$$

On integration by parts this becomes

$$\begin{aligned}(-1)^{n-r}\mu(\mu - 1) \ldots (\mu + r - n + 2) \Big[\Delta_C (t - z)^{\mu+r-n+1} q(t) f(t) \\ - \int_C (t - z)^{\mu+r-n+1} \frac{\mathrm{d}}{\mathrm{d}t}\{q(t)f(t)\}\,\mathrm{d}t\Big],\end{aligned} \tag{2.7.4}$$

and on integrating again we have

$$\begin{aligned}(-1)^{n-r}\mu(\mu - 1) \ldots (\mu + r - n + 3)\Big(\Delta_C \Big[(\mu + r - n + 2) q(t) f(t) \\ - (t - z)\frac{\mathrm{d}}{\mathrm{d}t}\Big\{q(t)f(t)\Big\}\Big](t - z)^{\mu+r-n+1} \\ - \int_C (t - z)^{\mu+r-n+2} \frac{\mathrm{d}^2}{\mathrm{d}t^2}\{q(t)f(t)\}dt\Big),\end{aligned} \tag{2.7.5}$$

and so on.

Consider then the equation

$$\sum_{k=0}^{n} a_k(z) w^{n-k}(z) = 0 \tag{2.7.6}$$

and let us suppose that the coefficients $a_k(z)$ can be expressed in terms of $m+1$ polynomials $q_0(z), q_1(z), \ldots q_m(z)$ by the relations

$$a_k(z) = (-1)^k \sum_{r=0}^{k} \binom{\mu+k-n}{k-r} q_r^{(k-r)}(z) \tag{2.7.7}$$

where for convenience we let $q_r(z) \equiv 0$ if $r > m$. The degree of $q_r(z)$ is at most $n - r$ and at least one of them must be of its full order. Then from (2.7.3) the equation is equivalent to

$$(-1)^n \int_C \left[\sum_{r=0}^{m} \mu(\mu-1)\ldots(\mu+r-n+1)(t-z)^{\mu+r-n} q_r(t) \right] \times f(t)\, \mathrm{d}t = 0. \tag{2.7.8}$$

If we integrate this by parts successively we have

$$(-1)^n \mu(\mu-1)\ldots(\mu+m+n-1)\Delta_C \Phi(z, t) + (-1)^n \mu(\mu-1)\ldots(\mu+m-n+1) \times \int_C (t-z)^{\mu+m-n} \left[\sum_{r=0}^{m} (-1)^m \frac{\mathrm{d}^{m-r}}{\mathrm{d}t^{m-r}} \{q_r(t) f(t)\} \right] \mathrm{d}t = 0$$

where

$$\Phi(z, t) = \sum_{r=0}^{m-1} \sum_{s=0}^{m-1-r} (-1)^s (\mu+m-n)\ldots \times (\mu-n+r+s+2)(t-z)^{\mu-n+r+s+1} \frac{\mathrm{d}^s}{\mathrm{d}t^s} \{q_r(t) f(t)\}.$$

Again we select the function $f(t)$ to satisfy the differential equation

$$\sum_{r=0}^{m} (-1)^m \frac{\mathrm{d}^{m-r}}{\mathrm{d}t^{m-r}} \{q_r(t) f(t)\} = 0, \tag{2.7.9}$$

and when $f(t)$ is known, we select C so that

$$\Delta_C \Phi(z, t) = 0. \tag{2.7.10}$$

We have therefore replaced an equation of order n by one of order m, and it is implicitly assumed that m is less than or equal to n. It is easy to see that if $a_k(z)$ itself is a polynomial of degree less than or equal to $n - k$, we can then express these coefficients in terms of n polynomials as above, and hence the order of the corresponding equation (2.7.9) is at most of order $n - 1$, and so is reduced.

2.8. A particular case of the Euler transformation

For the sake of simplicity we consider in detail only the case when the coefficients can be expressed in terms of two polynomials $q_0(z)$ and $q_1(z)$ of respective orders less than or equal to n and $n - 1$. This, therefore, includes all second-order equations wherein the coefficient $a_0(z)$ is quadratic, $a_1(z)$ is a linear function and a_2 is constant. Thus we assume that

$$a_0(z) = q_0(z)$$

and for $k \geqslant 1$

$$(-1)^k a_k(z) = \binom{\mu + k - n}{k} q_0^{(k)}(z) + \binom{\mu + k - n}{k - 1} q_1^{(k-1)}(z).$$

Then (2.7.9) reduces to

$$\frac{\mathrm{d}}{\mathrm{d}t}\{q_0(t)f(t)\} = q_1(t)f(t) \tag{2.8.1}$$

so that

$$q_0(t)f(t) = \exp\left[\int_{t_0}^{t} \{q_1(\tau)/q_0(\tau)\}\,\mathrm{d}\tau\right]. \tag{2.8.2}$$

The corresponding condition on the contours is that

$$\Delta_C(t - z)^{\mu - n + 1} q_0(t)f(t) = 0. \tag{2.8.3}$$

Let us consider the various possibilities. If $q_0(t)$ has its full degree n, then the degree of $q_1(t)$ is at most $n - 1$, and so $q_1(t)/q_0(t)$ can be expanded in a set of n partial fractions. Suppose first that all the roots of $q_0(t)$ are simple, say $\alpha_1, \alpha_2, \ldots \alpha_n$. Then

$$q_0(t)f(t) = \prod_{r=1}^{n} (t - \alpha_r)^{\beta_r} \tag{2.8.4}$$

for some set of constants β_r. Now from the logarithmic derivatives of (2.8.4)

$$\sum \beta_r = \lim_{t \to \infty} t q_1(t)/q_0(t) = B, \text{ say}$$

and so unless $B + \mu - n + 1 < 0$ we cannot let the contour go to infinity. We shall show, however, in the next section that, given two algebraic branch points, we can always choose a closed contour round them so that the function returns to its initial value. In this case we can therefore pair the branch point at $t = z$ with each of $\alpha_1, \alpha_2, \ldots \alpha_n$ respectively and derive n independent solutions.

If $q_0(t)$ has a multiple root, then correspondingly we get partial fractions, say,

$$\frac{b_N}{(t - \alpha)^N} + \frac{b_{N-1}}{(t - \alpha)^{N-1}} + \ldots + \frac{b_1}{(t - \alpha)},$$

leading to a factor of $q_0(t)f(t)$ of the type

$$(t-\alpha)^{b_1}\exp\left\{-\frac{b_N}{(N-1)(t-\alpha)^{N-1}}-\frac{b_{N-2}}{(N-2)(t-\alpha)^{N-2}}\cdots-\frac{b_2}{(t-\alpha)}\right\}.$$

There are therefore, as we have seen before, $N-1$ sectors of approach to $t=\alpha$ such that the exponential term will vanish. Correspondingly we can find $N-1$ contours starting and finishing at $t=\alpha$, and we get an additional contour from pairing the branch points at $t=z$ and $t=\alpha$. Thus once again we can find N independent integrals.

If the degree of $q_1(t)$ is equal to or exceeds $q_0(t)$, then $q_0(t)f(t)$ will contain a factor

$$\exp\left\{C_0t+\frac{C_1}{2}t^2+\ldots+\frac{C_{N-1}}{N}t^N\right\}$$

if the excess in the degree of $q_1(t)$ is $N-1$. Then there will be N sectors of approach to infinity such that the exponential factor tends to zero along it, and we may obtain N independent integrals from contours starting from infinity in one of these directions, encircling a branch point and returning to infinity. Thus once again we can obtain N independent integrals.

2.9 The contour for two distinct branch points

Consideration of contour integrals of either the Laplace or the Euler type leads to the possibility of solutions of the basic form

$$\int_C (t-\alpha_1)^{\beta_1-1}(t-\alpha_2)^{\beta_2-1}\phi_1(t)\,\mathrm{d}t \tag{2.9.1}$$

combined with a condition

$$\Delta_C(t-\alpha_1)^{\beta_1}(t-\alpha_2)^{\beta_2}\phi_2(t)=0, \tag{2.9.2}$$

where $\phi_1(t)$ and $\phi_2(t)$ are regular functions of t in a domain which includes the two branch points α_1 and α_2 and the whole of the contour C.

Then if neither β_1 nor β_2 is an integer we can choose for C the double-dumb-bell contour shown in Fig. 2.6. We may verify that this satisfies

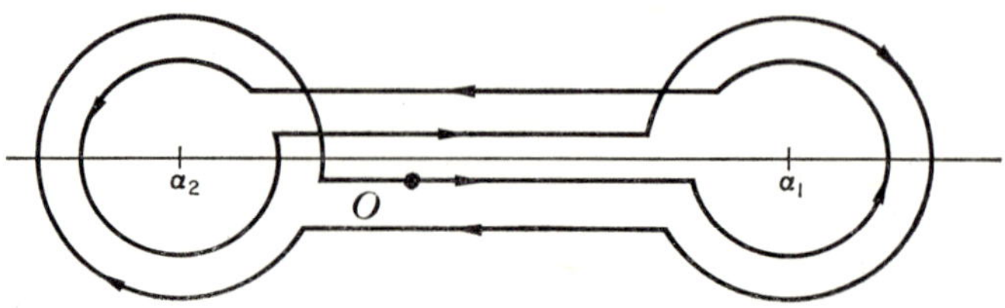

Fig. 2.6.

equation (2.9.2). Since $\phi_2(t)$ is regular in a region which includes C, it is single-valued on C. Let the argument of $(t - \alpha_1)^{\beta_1}(t - \alpha_2)^{\beta_2}\phi_2(t)$ at the point O be γ. The contour goes first round α_1 in the positive direction, changing this argument to $\gamma + 2\pi\beta_1$, then round α_2 in the positive direction, changing it to $\gamma + 2\pi(\beta_1 + \beta_2)$, back round α_1 in the negative direction, changing it to $\gamma + 2\pi\beta_2$, and finally round α_2 in the negative direction, when the argument returns to its initial value. Thus the total changes round C in both the argument and modulus are zero.

Let I_1 and I_2 be the values of integrals, with the same integrand as (2.9.1), but round simple closed curves, both passing through the point O, and round the points α_1 and α_2 respectively. Then the value of (2.9.1) is

$$\{1 - \exp(2\pi i\beta_2)\}I_1 - \{1 - \exp(2\pi i\beta_1)\}I_2, \tag{2.9.3}$$

which is clearly non-trivial if neither β_1 nor β_2 is an integer.

We can obtain the corresponding result if one of them is an integer by a limiting process. Put $\beta_2 = k + \varepsilon$ and consider

$$\lim_{\varepsilon\to 0} \frac{1 - e^{2\pi i\varepsilon}}{\varepsilon} I_1 - \{1 - \exp(2\pi i\beta_1)\}\frac{I_2}{\varepsilon}$$

or

$$-2\pi i I_1 - \{1 - \exp(2\pi i\beta_1)\}\left(\frac{\partial I_2}{\partial \beta_2}\right)_{\beta_2=k}. \tag{2.9.4}$$

In the usual way the latter term is

$$\int_{C_2} (t - \alpha_1)^{\beta_1-1}(t - \alpha_2)^{k-1} \log(t - \alpha_2)\phi_1(t)\, dt. \tag{2.9.5}$$

2.10 Illustrations of the Euler integral

ILLUSTRATION 6

The hypergeometric equation is

$$z(1 - z)w'' + \{c - (a + b + 1)z\}w' - abw = 0. \tag{2.10.1}$$

Then from (2.7.7) we have

$$q_0(z) = z(1 - z),$$
$$(\mu - 1)q_0'(z) + q_1(z) = (a + b + 1)z - c,$$
$$\tfrac{1}{2}\mu(\mu - 1)q_0''(z) + q_1'(z) = -ab,$$

and so

$$\mu = -a,$$
$$q_1(z) = (b - a - 1)z + (a + 1 - c). \tag{2.10.2}$$

Thus

$$\frac{q_1(t)}{q_0(t)} = \frac{b - c}{1 - t} + \frac{a + 1 - c}{t}$$

and so

$$w(z) = \int_C t^{a-c}(t-1)^{c-b-1}(t-z)^{-a}\,dt, \tag{2.10.3}$$

provided that

$$\Delta_C t^{1+a-c}(t-1)^{c-b}(t-z)^{-a-1} = 0. \tag{2.10.4}$$

Thus we derive a contour of the form of section 2.8 from any pair of the three branch points at 0, 1 and z. In particular, the contour round $t = 0$ and $t = 1$ can be contracted into the line joining them, and the integral will be a multiple of

$$\int_0^1 t^{a-c}(1-t)^{c-b-1}(t-z)^{-a}\,dt \tag{2.10.5}$$

if $\mathscr{R}(a-c+1) > 0$ and $\mathscr{R}(c-b) > 0$.

If $\mathscr{R}(b) > 0$, we can take a contour extending to infinity, but if $\mathscr{R}(c) > \mathscr{R}(b) > 0$ it would be sufficient to use

$$w(z) = \int_1^\infty t^{a-c}(t-1)^{c-b-1}(t-z)^{-a}\,dt. \tag{2.10.6}$$

If we write $t\tau = 1$, this becomes

$$w(z) = \int_0^1 \tau^{b-1}(1-\tau)^{c-b-1}(1-\tau z)^{-a}\,d\tau, \tag{2.10.7}$$

which in the usual notation (see Chapter 6) is

$$\frac{\Gamma(b)\,\Gamma(c-b)}{\Gamma(c)}F(a, b; c; z). \tag{2.10.8}$$

Illustration 7

The Legendre equation is

$$(1-z^2)w'' - 2zw' + \nu(\nu+1)w = 0. \tag{2.10.9}$$

In this case

$$\mu = -(\nu+1), \quad q_0(z) = 1 - z^2, \quad \text{and} \quad q_1(z) = -2(\nu+1)z.$$

Then

$$w(z) = \int_C \{(t^2-1)^\nu/(t-z)^{\nu+1}\}\,dt, \tag{2.10.10}$$

provided

$$\Delta_C(t^2-1)^{\nu+1}(t-z)^{-\nu-2} = 0. \tag{2.10.11}$$

In this case the choice of contours is simplified by the occurrence of ν linearly in the exponent of each of the branch points at $t = \pm 1$ and $t = z$. It is therefore possible to choose a simple closed curve round $t = 1$ and $t = z$, or round $t = -1$ and $t = z$, or a figure of eight going once round $t = 1$ and $t = -1$ in opposite directions. The first choice leads essentially to $P_\nu(z)$ and the third to $Q_\nu(z)$ (see Chapter 8). It would only be possible to take a contour extending to infinity if $\mathscr{R}(\nu) < 0$.

ILLUSTRATION 8

The equation

$$z^2w'' + (b - 2az)w' + a(a + 1)w = 0 \qquad (2.10.12)$$

leads to

$$\mu = a, \quad q_0(z) = z^2, \quad q_1(z) = 2z - b,$$

and hence to a solution

$$w(z) = \int_C \exp(b/t)(t - z)^a \, dt, \qquad (2.10.13)$$

provided

$$\Delta_C t^2 \exp(b/t)(t - z)^{a-1} = 0. \qquad (2.10.14)$$

Since there is no branch point at the origin, we can derive one integral as the limit of the double dumb-bell as the exponent tends to an integer, as in the final paragraph of section 2.9. To obtain a second solution we must use the strong singularity of exp (b/t). If we assume b is real and positive for simplicity, then the expression in (2.10.14) will tend exponentially to zero as the origin is approached by the negative real axis. Hence formally we would take a closed contour in the neighbourhood of the origin, and cusped at the origin, with its double tangent along the negative real axis. In fact our expression is single-valued near the origin, and so the contributions from the double tangent will cancel each other. Thus we may distort it into a closed curve round the origin.

ILLUSTRATION 9

As a final illustration let us consider again the confluent hypergeometric equation

$$zw'' + (c - z)w' - aw = 0. \qquad (2.10.15)$$

This leads to

$$\mu = -a, \quad q_0(z) = z, \quad q_1(z) = z + a + 1 - c$$

and hence to a solution

$$w(z) = \int_C e^t t^{a-c}(t - z)^{-a} \, dt, \qquad (2.10.16)$$

provided

$$\Delta_C \, e^t t^{a-c+1}(t - z)^{-a-1} = 0. \qquad (2.10.17)$$

In fact it is reasonably easy to see that this does not give any essentially new result, as the transformation $t = z\tau$ reduces (2.10.16) to a contour integral of the Laplace type.

EXERCISES

1. Find solutions in the form of contour integrals of Laplace type of
$$zw'' - 2w' - zw = 0$$
and hence deduce that a pair of independent solutions are
$$w_1(z) = (1 + z)e^{-z} \quad \text{and} \quad w_2(z) = (1 - z)e^{z}.$$
2. Find a pair of independent solutions of the Laplace type of
$$zw'' + \{(\alpha + \beta) - (a + b)z\}w' + \{abz - (a\beta - b\alpha)\}w = 0,$$
where a, b, α and β are real parameters. Consider the possible contours as α and β vary, keeping a and b fixed.
3. Find a pair of independent solutions of the Laplace type of
$$w'' + 2(z - a)w' + a^2w = 0,$$
where a is a real parameter.
4. Use the method of solution by contour integral to show that the coefficient of t in the Laurent expansion round $t = 0$ of $\exp(zt - 1/t)$ is a solution of
$$zw'' + w = 0.$$
Find also an independent second solution if $\mathscr{R}(z) > 0$.
5. Find a general solution of the equation
$$2zw''' + 5w'' - 2zw' - 3w = 0.$$
6. Find two independent solutions of the form
$$w(z) = \int_C f(t) \exp(tz^2)\, dt$$
of the equation
$$w'' + 4(1 + z^2)w = 0.$$
7. Find solutions in the form of contour integrals of the Euler type of
$$4(1 + z^2)w'' + 8zw' + 9w = 0.$$
8. Find contour integral solutions of the Euler type of
$$4z^2w'' + 4w' + w = 0.$$
9. Find contour-integral solutions of the Euler type of the Hermite differential equation of integer order n
$$w'' - 2zw' + 2nw = 0.$$

ANSWERS

2. $w = \int_C e^{zt}(t - a)^{\alpha-1}(t - b)^{\beta-1}\, dt.$

Representative contours are:

(i) $\alpha > 0, \beta > 0$; a line joining a to b and a line joining either a or b to ∞;

(ii) Either $\alpha > 0$ or $\beta > 0$ (let us suppose $\alpha > 0$); a curve round $t = b$, starting and finishing at $t = a$, and a line joining $t = a$ to ∞.

(iii) $\alpha \leqslant 0$, $\beta \leqslant 0$; two curves each starting and finishing at ∞, one round $t = a$ and the other round $t = b$ (which will degenerate into closed curves round $t = a$ or $t = b$ if α or β is an integer).

In every case infinity is to be approached so that $\mathscr{R}(zt) < 0$.

3. $$w_1(z) = \int_0^\infty \exp(-u^2/4)u^{a^2-1} \cos\{(z-a)u\}\, du$$

$$w_2(z) = \int_0^\infty \exp(-u^2/4)u^{a^2-1} \sin\{(z-a)u\}\, du.$$

4. $$w_2(z) = \int_C t^{-2} \exp(zt - t^{-1})\, dt,$$

where C is a curve leaving the origin along the positive real axis and going to infinity on the negative real axis.

5. $$w(z) = \sum_{s=1}^{3} A_s w_s(z),$$

where the A_s are constants, $w_s(z) = \int_a^b e^{zt}t^{1/2}(t^2-1)^{-1/2}\, dt$, where if $\mathscr{R}(z) > 0$, the pair (a, b) are for $w_1(z)$, $(-\infty, -1)$, for $w_2(z)$, $(-1, 0)$ and for $w_3(z)$, $(0, 1)$. Corresponding lines are to be chosen if $\mathscr{R}(z) \leqslant 0$.

6. $$w_1(z) = \int_{-1}^{1} (1-t^2)^{-3/4} \exp\{i(z^2t + \tanh^{-1}t)\}\, dt$$

$$w_2(z) = \int_1^\infty (t^2-1)^{-3/4} \exp\{i(z^2t + \tanh^{-1}t)\}\, dt.$$

7. $$w_1(z) = \int_{-1}^{1} (t + iz)^{3/2}(1-t^2)^{-1/2}\, dt$$

$$w_2(z) = \int_1^{-iz} (t + iz)^{3/2}(t^2-1)^{-1/2}\, dt.$$

8. $$w(z) = \int_C t^{-1}(t-z)^{1/2} \exp(-1/t)\, dt.$$

Two suitable contours would be (i) a closed curve encircling $t = z$, and cusped at $t = 0$ with the positive real axis as semi-tangent; (ii) the double dumb-bell round $t = z$ and $t = 0$, in the degenerate case of an integer exponent at $t = 0$.

9. $$w(z) = \int_C (t-z)^{-n} \exp(-t^2)\, dt.$$

Two contours are: (i) a closed curve round $t = z$; (ii) the real axis $(-\infty, \infty)$ indented to avoid $t = z$ if z is real.

CHAPTER 3

Oscillation Theorems and Sturm–Liouville Theory

3.1 The adjoint equation

In those parts of mathematical physics which are governed by linear partial differential equations, and these include matters of considerable interest, it is frequently possible to obtain elementary solutions of the equations by the method of 'separation of variables' (see Appendices I and II). We shall be primarily concerned in the later chapters of this book with functions which are the solutions of differential equations obtained in this way. These functions are often of an oscillatory nature and the distribution and separation of the zeros is a matter of some importance. Another general property which is of paramount concern is that of the 'completeness' of the set of solutions of these equations, obtained by varying a parameter of the equations. This determines whether we are able to use the linear property of the governing partial differential equations to build up from the elementary solutions a more general solution which will satisfy the particular requirements of the boundary and initial conditions. This chapter is concerned with the deductions which can be made from the differential equations in regard to these two properties.

Let us suppose that we are concerned with the solution of the differential equation.

$$L(y) \equiv \sum_{k=0}^{n} a_k(x)\frac{d^{n-k}y}{dx^{n-k}} = 0 \tag{3.1.1}$$

(N.B. in this chapter we shall assume throughout that the independent variable is real) on an interval $[a, b]$. Let $u(x)$ be a function of x which is continuously differentiable n times on $[a, b]$. Then integration by parts

$n - k$ times gives

$$\int_a^b u(x)a_k(x)\frac{\mathrm{d}^{n-k}y}{\mathrm{d}x^{n-k}}\mathrm{d}x = \sum_{j=0}^{n-k-1}(-1)^j\Delta_a{}^b\frac{\mathrm{d}^j}{\mathrm{d}x^j}\{u(x)a_k(x)\}\frac{\mathrm{d}^{n-k+1-j}y}{\mathrm{d}x^{n-k+1-j}}$$
$$+ (-1)^{n-k}\int_a^b \frac{\mathrm{d}^{n-k}}{\mathrm{d}x^{n-k}}\{a_k(x)u(x)\}y(x)\,\mathrm{d}x,$$

where $\Delta_a{}^b$ is the difference of the function between $x = a$ and $x = b$.

Thus associated with equation (3.1.1) is a second linear equation $M(u) = 0$, which is so constructed that

$$\{uL(y) - yM(u)\}$$

is the derivative of a bilinear form in y, u and their derivatives. We can therefore see that this defines the differential operator M uniquely and that

$$M(u) \equiv \sum_{k=0}^{n}(-1)^{n-k}\frac{\mathrm{d}^{n-k}}{\mathrm{d}x^{n-k}}\{a_k(x)u\}. \tag{3.1.2}$$

The operator M is said to be the adjoint of the differential operator L, and the corresponding equation $M(u) = 0$ is the adjoint equation.

It may happen that an equation is its own adjoint. In the case of the second-order equation

$$L(y) \equiv \sum_{k=0}^{2}a_k(x)\frac{\mathrm{d}^{n-k}y}{\mathrm{d}x^{n-k}} = 0$$

the adjoint equation is

$$M(u) \equiv a_0(x)\frac{\mathrm{d}^2u}{\mathrm{d}x^2} + \{2a_0'(x) - a_1(x)\}\frac{\mathrm{d}u}{\mathrm{d}x}$$
$$+ \{a_0''(x) - a_1'(x) + a_2(x)\}u = 0 \tag{3.1.3}$$

and so $L(y)$ is its own adjoint if and only if

$$a_0'(x) = a_1(x). \tag{3.1.4}$$

If this condition is satisfied, the equation is said to be symmetric. A symmetric equation with boundary conditions of a certain type discussed below is called a self-adjoint system.

In this chapter we shall consider symmetric differential operators of second order. It is worth remarking that any equation of second order can be reduced to a symmetric form on any interval on which the coefficients are continuous, $a_0'(x)$ exists almost everywhere, and $a_0(x) \neq 0$.

For the equation

$$a_0(x)\frac{\mathrm{d}^2y}{\mathrm{d}x^2} + a_1(x)\frac{\mathrm{d}y}{\mathrm{d}x} + a_2(x)y = 0, \tag{3.1.5}$$

we multiply through by $f(x)$ and choose $f(x)$ so that

$$f(x)a_1(x) = \frac{\mathrm{d}}{\mathrm{d}x}\{a_0(x)f(x)\}. \tag{3.1.6}$$

In fact

$$a_0(x)f(x) = \exp\left[\int_{x_0}^{x} \{a_1(t)/a_0(t)\}\,\mathrm{d}t\right]. \tag{3.1.7}$$

We therefore take for the standard form of our equation

$$\frac{\mathrm{d}}{\mathrm{d}x}\left\{p(x)\frac{\mathrm{d}y}{\mathrm{d}x}\right\} + q(x)y = 0. \tag{3.1.8}$$

If $p(x) \neq 0$ on the interval considered, we may put

$$\mathrm{d}\xi = \{p(x)\}^{-1}\,\mathrm{d}x, \tag{3.1.9}$$

and the equation will reduce to

$$\frac{\mathrm{d}^2y}{\mathrm{d}\xi^2} + Q(\xi)y = 0 \tag{3.1.10}$$

where

$$Q(\xi) = p(x)q(x). \tag{3.1.11}$$

3.2 The oscillation theorems

It is assumed in the following work that $p'(x)$ and $q(x)$ are continuous on any interval considered. Then the first result we may prove is the following.

If on the interval $[a, b]$, $p(x) > 0$ and $q(x) \leqslant 0$, then any solution of the equation (3.1.8) either vanishes identically on the closed interval, or has at most one zero in the interval. In fact we prove slightly more; either $y(x)$ vanishes identically or $y(x)y'(x)$ has not more than a single zero in $[a, b]$.

We construct the function

$$\phi(x) = p(x)y(x)y'(x). \tag{3.2.1}$$

Then

$$\begin{aligned}\phi'(x) &= y(x)\frac{\mathrm{d}}{\mathrm{d}x}\{p(x)y'(x)\} + p(x)\{y'(x)\}^2 \\ &= p(x)\{y'(x)\}^2 - q(x)\{y(x)\}^2 \geqslant 0,\end{aligned} \tag{3.2.2}$$

and the equality sign can hold only if $y'(x) = 0$ and either $q(x) = 0$ or $y(x) = 0$. If $q(x)$ is not identically zero, either $y'(x) = y(x) = 0$ at every point of the interval, or $\phi'(x)$ is positive at least at some points of the interval. Then $\phi(x)$ is an increasing function and so $\phi(x)$ has at most a single zero on $[a, b]$. Since $p(x) > 0$, this implies that $y(x)y'(x)$ has a single zero on $[a, b]$. If $q(x) \equiv 0$, the same result follows unless $y'(x) = 0$

on $[a, b]$, when clearly the solution is a constant, and so is either zero everywhere or has no zero on $[a, b]$.

In any interval on which $p(x) \neq 0$, no solution of (3.1.8) can have a double zero, unless it vanishes identically. For at a double zero x_0 we have $y(x_0) = y'(x_0) = 0$. The trivial solution $y(x) \equiv 0$ satisfies both the equation and these two boundary conditions, and so by the uniqueness theorem for ordinary differential equations this is the only solution [7].

3.3 The comparison theorems

We now establish the first comparison theorem. We suppose that we have two differential equations of the form (3.1.8)

$$\frac{\mathrm{d}}{\mathrm{d}x}\left\{p_1(x)\frac{\mathrm{d}y}{\mathrm{d}x}\right\} + q_1(x)y = 0 \tag{3.3.1}$$

and

$$\frac{\mathrm{d}}{\mathrm{d}x}\left\{p_2(x)\frac{\mathrm{d}u}{\mathrm{d}x}\right\} + q_2(x)u = 0 \tag{3.3.2}$$

and $p_1(x) \geqslant p_2(x) > 0$ and $q_2(x) \geqslant q_1(x)$ on $[a, b]$, then there is at least one zero of $u(x)$ between any pair of consecutive zeros of $y(x)$. This result does not depend on which solutions of (3.3.1) and (3.3.2) are taken.

The result follows from Picone's identity, which we now establish. Consider

$$\psi(x) = y(x)\{p_1(x)y'(x)u(x) - p_2(x)y(x)u'(x)\}\{u(x)\}^{-1}.$$

Then

$$\begin{aligned}\frac{\mathrm{d}\psi}{\mathrm{d}x} &= y(x)\frac{\mathrm{d}}{\mathrm{d}x}\{p_1(x)y'(x)\} + p_1(x)\{y'(x)\}^2 - \{y(x)\}^2\{u(x)\}^{-1}\frac{\mathrm{d}}{\mathrm{d}x}\{p_2(x)u'(x)\}\\ &\qquad - 2p_2(x)y(x)y'(x)u'(x)\{u(x)\}^{-1} + p_2(x)\{y(x)\}^2\{u'(x)/u(x)\}^2\\ &= \{q_2(x) - q_1(x)\}\{y(x)\}^2 + \{p_1(x) - p_2(x)\}\{y'(x)\}^2\\ &\qquad + p_2(x)\{u(x)y'(x) - u'(x)y(x)\}^2\{u(x)\}^{-2}. \end{aligned} \tag{3.3.3}$$

Thus $\mathrm{d}\psi/\mathrm{d}x \geqslant 0$ and so $\psi(x)$ is non-decreasing in the interval.

Suppose then that x_1 and x_2 are consecutive zeros of $y(x)$ with $a < x_1 < x_2 < b$ and suppose $u(x)$ is not zero for any x in $x_1 \leqslant x \leqslant x_2$. We may assume without loss of generality that $y(x) > 0$ in $x_1 < x < x_2$ and $u(x) > 0$ in $x_1 \leqslant x \leqslant x_2$, since otherwise we may replace either by its negative value.

Then $\psi(x_1) = 0$, and so if $\psi'(x)$ is identically zero, $\psi(x)$ is zero on the whole interval, or if $\psi'(x) > 0$, then $\psi(x)$ increases from zero and so takes positive values. By assumption, however, $\psi(x_2) = 0$. If we then exclude the case $\psi(x) \equiv 0$, $\psi(x)$ must return to zero. However, this is a contradiction as long as $\psi(x)$ remains finite. Thus $\psi(x)$ must go off to infinity at some point x_0 within the interval. Both $y(x)$ and $p_1(x)y'(x)u(x) - p_2(x)y(x)u'(x)$ remain finite, and so it follows that $\{u(x)\}^{-1}$ is unbounded in the interval. Thus $u(x)$ must have a zero at x_0, which is a contradiction.

Since this theorem is applicable to the case when $p_1(x) = p_2(x)$ and $q_1(x) = q_2(x)$, we may deduce that if $y_1(x)$ and $y_2(x)$ are independent solutions of (3.1.8), then between each pair of zeros of $y_1(x)$ there is a zero of $y_2(x)$ and vice versa. Thus the zeros of the two solutions must interlace with one another.

We may make a further deduction from the Picone formula. We shall assume that $p_1(x)$, $p_2(x)$, $q_1(x)$ and $q_2(x)$ satisfy the same conditions as before. Suppose we pick solutions of (3.3.1) and (3.3.2), both of which vanish at x_1 and have the same gradient at this point, say, $y'(x_1) = u'(x_1) = m > 0$. The point x_1 is an ordinary point of the differential equation, and so $y(x)$ and $u(x)$ are analytic near x_1. Then

$$y(x) = m(x - x_1) + O\{(x - x_1)^2\}$$
$$u(x) = m(x - x_1) + O\{(x - x_1)^2\}$$

and so

$$\psi(x_1) = 0$$

and

$$\psi'(x_1) = m^2\{p_1(x_1) - p_2(x_1)\} \geqslant 0.$$

Again we deduce by a similar argument that if $y(x_2) = 0$, then $u(x_0) = 0$ for some x_0 in (x_1, x_2). Hence if the solutions of the equation are oscillatory, then $u(x)$ oscillates 'faster' than $y(x)$.

If the equations compared have $p_1(x) = p_2(x)$, but $q_2(x) \geqslant q_1(x)$, the Picone formula may be replaced by the simpler identity

$$\frac{\mathrm{d}}{\mathrm{d}x}[p(x)\{y'(x)u(x) - u'(x)y(x)\}] = \{q_2(x) - q_1(x)\}uy. \qquad (3.3.4)$$

3.4 The distribution of the zeros

Let us apply the first comparison theorem to the equation (3.1.10). Suppose that $M \geqslant Q(\xi) \geqslant m > 0$ on $[a, b]$. Select

$$p_1(\xi) = p_2(\xi) = 1, \quad q_1(\xi) = Q(\xi), \quad q_2(\xi) = M,$$

and so by the theorem there is at least one zero of any solution of

$$\frac{\mathrm{d}^2u}{\mathrm{d}\xi^2} + Mu = 0 \qquad (3.4.1)$$

between any pair of consecutive zeros of any solution of (3.1.10). The general solution of (3.4.1) is

$$u(\xi) = A\cos(\xi\sqrt{M} + \varepsilon) \qquad (3.4.2)$$

where A and ε are constants of integration. All these zeros are spaced $\pi M^{-1/2}$ apart, though their absolute position can be shifted by changing ε. Thus if ξ_1, ξ_2 are consecutive zeros of any solution of (3.1.10) we must have

$$\pi M^{-1/2} \leqslant \xi_1 - \xi_0, \tag{3.4.3}$$

for if not it would be possible to adjust ε so that no zero of $u(x)$ fell within $\xi_0 < \xi < \xi_1$, contrary to the theorem.

Similarly, comparing (3.1.10) and

$$\frac{\mathrm{d}^2 v}{\mathrm{d}x^2} + mv = 0, \tag{3.4.4}$$

we deduce that

$$\pi m^{-1/2} \geqslant \xi_1 - \xi_0. \tag{3.4.5}$$

Since these give upper and lower bounds for the spacing of the zeros of (3.1.10), we may deduce that if n is the number of zeros of $y(\xi)$ in $[a, b]$, then

$$(b - a)m^{1/2} - \pi \leqslant n\pi \leqslant (b - a)M^{1/2} + \pi. \tag{3.4.6}$$

3.5 Sturm–Liouville systems

It was mentioned above that in the definition of a self-adjoint differential system some restrictions are imposed on the boundary conditions to be satisfied, in addition to the requirement that the equation is symmetric.

It is known that a second-order equation requires two boundary conditions. If two independent conditions are specified at the same point, this, in conjunction with the equation, is equivalent to giving the value of the solution and its first derivative at the point, provided that the coefficient $p(x) \neq 0$ there. These are the conditions under which uniqueness is proved, and so we will not consider this case.

If we consider solutions of (3.1.8) on a finite interval on which $p(x) \neq 0$, all solutions are bounded, since every point is an ordinary point of the equation. Hence the obvious alternative is to give 'two-point' boundary conditions. These in general take the form

$$h_{11}y'(a) + h_{12}y(a) + k_{11}y'(b) + k_{12}y(b) = 0, \tag{3.5.1}$$

and

$$h_{21}y'(a) + h_{22}y(a) + k_{21}y'(b) + k_{22}y(b) = 0. \tag{3.5.2}$$

Clearly the condition that these are independent boundary conditions is that the matrix

$$\begin{pmatrix} h_{11} & h_{12} & k_{11} & k_{12} \\ h_{21} & h_{22} & k_{21} & k_{22} \end{pmatrix}$$

has rank 2. It is shown in [7] that we need to impose the restriction

$$Kp(a) = Hp(b), \tag{3.5.3}$$

where

$$H = \begin{vmatrix} h_{11} & h_{12} \\ h_{21} & h_{22} \end{vmatrix} \quad \text{and} \quad K = \begin{vmatrix} k_{11} & k_{12} \\ k_{21} & k_{22} \end{vmatrix}$$

if we are to preserve the full symmetry of the system.

We restrict self-adjoint systems to symmetric equations satisfying this restriction. Included within this class of system are those with the boundary conditions

$$h_1 y'(a) + h_2 y(a) = 0, \tag{3.5.4}$$

and

$$k_1 y'(b) + k_2 y(b) = 0. \tag{3.5.5}$$

The 'periodic' boundary conditions

$$y(a) = y(b), \quad y'(a) = y'(b) \tag{3.5.6}$$

are included if and only if $p(a) = p(b)$, since in this case $H = K = -1$.

We shall consider differential equations of Sturm–Liouville type. These have the form

$$\frac{\mathrm{d}}{\mathrm{d}x}\left\{p(x)\frac{\mathrm{d}y}{\mathrm{d}x}\right\} + \{q(x) + \lambda r(x)\}y = 0, \tag{3.5.7}$$

where λ is a variable parameter. Most of the well-known equations of mathematical physics are of this form, as Table 3.1 shows.

TABLE 3.1

Equation	$p(x)$	$q(x)$	$r(x)$	Usual range	Regular or non-regular
Harmonic	1	—	1	$[a, b]$	Regular
Harmonic	1	—	1	$[0, \infty]$	Non-regular
Legendre	$1 - x^2$	—	1	$[-1, 1]$	Non-regular
Associated Legendre	$1 - x^2$	$-\dfrac{m^2}{1 - x^2}$	1	$[-1, 1]$	Non-regular
Bessel	x	$-\dfrac{\nu^2}{x}$	x	$[0, a]$ or $[0, \infty]$	Non-regular
Laguerre	$x^\alpha e^{-x}$	—	$x^\alpha e^{-x}$	$[0, \infty]$	Non-regular
Hermite	$\exp(-x^2)$	—	$2\exp(-x^2)$	$[-\infty, \infty]$	Non-regular

If we have boundary conditions like (3.5.4) and (3.5.5) and $p(x) > 0$, $r(x) > 0$ on $[a, b]$, the differential system specified by (3.5.7) with (3.5.4) and (3.5.5) is said to be regular.

In the regular case we intuitively expect to find that a solution is possible only for a restricted set of values of the parameter λ. A general solution of (3.5.7) would have the form

$$y(x, \lambda) = Ay_1(x, \lambda) + By_2(x, \lambda) \tag{3.5.8}$$

where $y_1(x, \lambda)$ and $y_2(x, \lambda)$ are a pair of independent solutions, and A and B are constants which might depend on λ. If this satisfies (3.5.4), then for some constant C

$$y(x, \lambda) = C[\{h_1 y_2'(a, \lambda) + h_2 y_2(a, \lambda)\}y_1(x, \lambda) - \{h_1 y_1'(a, \lambda) + h_2 y_1(a, \lambda)\}y_2(x, \lambda)] \tag{3.5.9}$$

and if (3.5.5) is satisfied,

$$\{h_1 y_2'(a, \lambda) + h_2 y_2(a, \lambda)\}\{k_1 y_1'(b, \lambda) + k_2 y_1(b, \lambda)\} = \{h_1 y_1'(a, \lambda) + h_2 y_1(a, \lambda)\}\{k_1 y_2'(b, \lambda) + k_2 y_2(b, \lambda)\}. \tag{3.5.10}$$

This can be interpreted as an equation in λ. We cannot, however, by this method derive general information about the distribution of the roots. In the next section we shall look at an elementary example, and then in the section following prove a more complete result.

Any value of λ for which a solution of the problem is possible is called an eigenvalue or characteristic value of the equation, and the corresponding solution (which as is seen in (3.5.9), will be known only up to a multiplying constant) is called the eigenfunction.

3.6 An eigenvalue problem for the harmonic equation

Consider the equation

$$\frac{d^2y}{dx^2} + \lambda y = 0 \tag{3.6.1}$$

with the boundary conditions (3.5.4) and (3.5.5). Since x does not occur explicitly in the coefficients of this equation, we can shift the origin. There is then no loss of generality in taking $a = 0$. For the sake of simplicity we shall also assume that $h_1 = 0$, $h_2 = 1$ in (3.5.4).

Thus we have three cases to consider:

(a) $\lambda < 0$ and $y(x) = A \sinh \{(-\lambda)^{1/2}x\}$; (3.6.2)

(b) $\lambda = 0$ and $y(x) = Ax$; (3.6.3)

(c) $\lambda > 0$ and $y(x) = A \sin (\lambda^{1/2}x)$. (3.6.4)

In case (a), therefore, we need to satisfy

$$(-\lambda)^{1/2}k_1 \cosh \{(-\lambda)^{1/2}b\} + k_2 \sinh \{(-\lambda)^{1/2}b\} = 0. \tag{3.6.5}$$

If for simplicity we write $\eta = (-\lambda)^{1/2}b$, and $k_2 bm + k_1 = 0$, this takes the simpler form

$$\tanh \eta = m\eta. \tag{3.6.6}$$

It is not difficult to show that all roots of this equation are real or purely imaginary. The latter possibility corresponds to case (c) and is dealt with below. Simple graphical considerations lead to the conclusion that if $m \leqslant 0$ or if $m > 1$, the only root is $\eta = 0$ or $\lambda = 0$; if $m = 1$, $\eta = 0$ is a double root; while if $0 < m < 1$, there is in addition one positive and one negative root. The eigenvalue $\lambda = 0$ is treated under case (c) and so we may assert that:

If $0 < k_2b/k_1 < -1$ there is just one real negative eigenvalue and the corresponding eigenfunction is shown in (3.6.2).

In case (b) $k_1 + k_2b = 0$, and if this is satisfied we have an eigenfunction of the form (3.6.3). Other eigenfunctions will be of the form discussed for case (c) below.

Finally in case (c) we need to satisfy

$$\lambda^{1/2}k_1 \cos(\lambda^{1/2}b) + k_2 \sin(\lambda^{1/2}b) = 0, \tag{3.6.7}$$

or if we define m as above, but now put $\eta = \lambda^{1/2}b$,

$$\tan \eta = m\eta. \tag{3.6.8}$$

Here again the roots which we need to consider are real, and for every m, $\eta = 0$ is a root and the other roots can be paired symmetrically into positive and negative. Consider then the positive roots. Once again this is most easily seen graphically. If $m < 0$, the first positive root lies between $\pi/2$ and π and there are an infinity of roots which tend asymptotically to $(n + \frac{1}{2})\pi$ from above. If $0 < m < 1$, the first positive root lies between π and $3\pi/2$; if $m = 1$, $\eta = 0$ is a double root and the next root lies between π and $3\pi/2$; and if $m > 1$, the first root lies between 0 and $\pi/2$. In every case for which $m > 0$ the roots tend asymptotically to $(n + \frac{1}{2})\pi$ from below.

The analogous problem with $h_1 \neq 0$ can be treated similarly. In this case also the location of the eigenvalues is most easily seen graphically. Instead, however, of the intersections of $\xi = \tan \eta$ or $\xi = \tanh \eta$ with the line $\xi = m\eta$, we consider the intersection with $\xi = m\eta/(1 - c\eta^2)$ for various values of m and c.

3.7 The regular eigenfunction problem

We shall prove that, if $p(x) > 0$ and $r(x) > 0$ on some interval which includes $[a, b]$, there is a countable infinity of eigenvalues for the system (3.5.7) coupled with (3.5.4) and (3.5.5), and the set $\lambda_1, \lambda_2, \ldots \lambda_n, \ldots$ has its only limit point at ∞.

Since $p(x) > 0$ we can apply the transformation (3.1.9), and the equation becomes

$$\frac{d^2y}{d\xi^2} + \{Q(\xi) + \lambda R(\xi)\}y = 0, \tag{3.7.1}$$

where $Q(\xi) = p(x)q(x)$ and $R(\xi) = p(x)r(x) > 0$.

Thus if M and m are the maximum and minimum of $Q(\xi) + \lambda R(\xi)$ on the ξ interval corresponding to $[a, b]$, then both m and M tend linearly to ∞ as $\lambda \rightarrow \infty$. (Note: since $p(x) > 0$, $[a, b]$ is mapped (1–1) on the corresponding interval of ξ.) It follows therefore from the estimate of (3.4.6) that whatever solution of (3.7.1) we consider, the number of zeros contained in $[a, b]$ increases as λ increases and for large λ is $O(\lambda^{1/2})$.

Let us consider solutions of (3.5.7) which satisfy (3.5.4). The parameter λ enters linearly into just one coefficient of the equation, and this coefficient varies continuously with λ. Then by the usual theorem [7] the solution of the equation varies continuously with λ.

Suppose, for some λ_*, this solution has a zero at x_0, where $\alpha < x_0 < \beta$. We shall assume that $p(x) > 0$, $r(x) > 0$ on $[\alpha, \beta]$ with $\alpha < a < b < \beta$. For reasons given earlier, x_0 is a simple zero. Hence we can find an $\varepsilon > 0$ so that both $x_0 - \varepsilon$ and $x_0 + \varepsilon$ lie in $[\alpha, \beta]$ and $y(x_0 - \varepsilon, \lambda_*)$ and $y(x_0 + \varepsilon, \lambda_*)$ have opposite signs, and $y(x, \lambda_*)$ vanishes once, at x_0, in the sub-interval. We may assume without loss of generality that $y(x_0 - \varepsilon, \lambda_*) < 0$ and $y(x_0 + \varepsilon, \lambda_*) > 0$.

Since the solutions vary continuously with λ, we can find a δ so that

$$y(x_0 - \varepsilon, \lambda) < 0 \quad \text{and} \quad y(x_0 + \varepsilon, \lambda) > 0$$

if $| \lambda - \lambda_* | < \delta$. Thus $y(x, \lambda)$ has a zero in $| x - x_0 | < \varepsilon$ for every $| \lambda - \lambda_* | < \delta$ and so the position of each zero is a continuous function of λ. As λ varies, no zero can coincide with a, because from (3.5.4) $y(a, \lambda) = y'(a, \lambda) = 0$ and hence $y(x, \lambda) \equiv 0$. Since, then, the number of zeros in $[a, b]$ increases as λ increases, and since each individual zero varies continuously with λ, they must enter the interval from the right. Thus there is a set of values μ_1, μ_2, μ_3, etc., of λ for which

$$y(b, \mu_j) = 0, \qquad j = 1, 2, 3, \text{etc.}, \tag{3.7.2}$$

and we may number these so that the solution with $\lambda = \mu_j$ vanishes $j - 1$ times in the interior of $[a, b]$. This proves the theorem in the case $k_1 = 0$, $k_2 \neq 0$.

Now suppose $y(x, \lambda)$ is a solution of the equation with parameter λ, and $u(x, \hat{\lambda})$ is a solution with parameter $\hat{\lambda}$, where $\hat{\lambda} > \lambda$ and both satisfy (3.5.4). Then by the Picone formula

$$\frac{d}{dx}[p(x)y(x, \lambda)\{u(x, \hat{\lambda})\}^{-1}\{y'(x, \lambda)u(x, \hat{\lambda}) - y(x, \lambda)u'(x \hat{\lambda})\}] > 0$$

and hence

$$\frac{y'(b,\lambda)}{y(b,\lambda)} > \frac{u'(b, \hat{\lambda})}{u(b,\hat{\lambda})}. \tag{3.7.3}$$

It follows that if λ and $\hat{\lambda}$ are both contained between a pair of zeros, say, μ_{j-1} and μ_j, keeping $\hat{\lambda}$ fixed and decreasing λ, $y'(b, \lambda)/y(b, \lambda)$ must increase and must tend to $+\infty$ as $\lambda \to \mu_{j-1}$. Similarly, on keeping λ fixed and increasing $\hat{\lambda}$, $u'(b, \hat{\lambda})/u(b, \hat{\lambda})$ decreases and must tend to $-\infty$ as $\hat{\lambda} \to \mu_j$.

Thus between each pair μ_{j-1} and μ_j, $y'(b, \lambda)/y(b, \lambda)$ is strictly decreasing and so takes the value $-k_2/k_1$ once in the interval. This completes the proof for the regular case.

Unfortunately even the standard functions of mathematical physics involve the non-regular case, as the table of Sturm–Liouville equations illustrates. Other than the harmonic equation, the only regular problem which is likely to appear in practice is the solution of the Bessel equation on a finite interval with $a > 0$ and boundary conditions like (3.5.4) or (3.5.5). This occurs in the solution of the Laplace equation inside an annular cylinder of internal radius a and external radius b.

The non-regular case is usually characterized by an infinite or semi-infinite interval or by the vanishing of $p(x)$ at one or both ends of the interval, or by a combination of these. That these two characteristics are related is shown, for example, by the Legendre equation.

$$\frac{d}{dx}\left\{(1 - x^2)\frac{dy}{dx}\right\} + \nu(\nu + 1)y = 0. \tag{3.7.4}$$

This is a Sturm–Liouville equation with $p(x) = 1 - x^2$ and so $p(x)$ vanishes at ± 1, and the system is non-regular on $[-1, 1]$. The transformation

$$x = \tanh \xi$$

reduces the equation to

$$\frac{d^2y}{d\xi^2} + \nu(\nu + 1) \sec^2 \xi y = 0 \tag{3.7.5}$$

and the interval $[-1, 1]$ is mapped into $[-\infty, \infty]$.

The eigenvalue problem for the non-regular case is beyond the scope of this book, and the reader is referred to either [15] or [4]. In these cases one or both of the boundary conditions is usually replaced by a condition of boundedness or integrability of the solution.

In a commonly occurring problem involving the Bessel equation, (3.5.4) is replaced by the condition that the solution is bounded as $x \to +0$, while the second condition remains like (3.5.5). In the case of the Legendre function, a boundedness condition is equivalent to two conditions (see Chapter 8) and so no further one is required. The condition that $y(x, \nu)$ is bounded as $x \to +1$ restricts the solution to multiples of the solution usually denoted by $P_\nu(x)$, and the condition that $P_\nu(x)$ is bounded as $x \to -1$ then implies that ν is an integer.

It is sufficient to say that in these cases there can be either a discrete set of eigenvalues, a continuous set of eigenvalues or a mixture of the two. The discrete set corresponds to the problems associated with Legendre, Bessel, Laguerre or Hermite functions. The continuous set corresponds to the integral transforms.

3.8 The orthogonality of eigenfunctions

Let us suppose that λ_m and λ_n are distinct eigenvalues of the equations (3.5.7) and the corresponding eigenfunctions are $y_m(x)$ and $y_n(x)$. Then we see from the equations that

$$y_n(x)\frac{\mathrm{d}}{\mathrm{d}x}\left\{p(x)\frac{\mathrm{d}y_m}{\mathrm{d}x}\right\} - y_m(x)\frac{\mathrm{d}}{\mathrm{d}x}\left\{p(x)\frac{\mathrm{d}y_n}{\mathrm{d}x}\right\} + (\lambda_m - \lambda_n)r(x)y_m(x)y_n(x) = 0. \quad (3.8.1)$$

Integration yields

$$p(b)\{y_n(b)y_m'(b) - y_m(b)y_n'(b)\} - p(a)\{y_n(a)y_m'(a) - y_m(a)y_n'(a)\} + (\lambda_m - \lambda_n)\int_a^b r(x)y_m(x)y_n(x)\,\mathrm{d}x = 0. \quad (3.8.2)$$

If the contribution from the first two terms vanishes we may deduce that

$$\int_a^b r(x)y_m(x)y_n(x)\,\mathrm{d}x = 0, \quad (3.8.3)$$

since we have assumed that $\lambda_n \neq \lambda_m$. In this case we say that the set of functions $\{y_n(x)\}$ is orthogonal on the interval $[a, b]$ with weight function $r(x)$, and that the set of functions collectively forms an orthogonal set.

If at a, both $y_m(x)$ and $y_n(x)$ satisfy the boundary condition (3.5.4), then on eliminating h_1 and h_2 between the two conditions we derive

$$y_n(a)y_m'(a) - y_m(a)y_n'(a) = 0. \quad (3.8.4)$$

Thus in the regular case when the eigenfunctions satisfy (3.5.7) with boundary conditions (3.5.4) and (3.5.5), they will certainly form an orthogonal set. We can deduce (3.8.3) under more general conditions. If at one end of the interval, say at a, $p(a) = 0$, and we impose the conditions that all solutions and their derivatives are finite there, then clearly

$$\lim_{x \to a} p(x)\{y_n(x)y_m'(x) - y_m(x)y_n'(x)\} = 0. \quad (3.8.5)$$

This applies to the Legendre equation when a condition of boundedness is imposed at each end of the interval. As we pointed out in the previous section, the usual eigenvalue problem for this equation is of this type. It is left as an exercise to the reader, when he has studied Chapter 8, to show that condition (3.8.5) is not satisfied for any Legendre functions at ± 1, unless these are the polynomial solutions.

If we have an infinite or semi-infinite interval it is clearly sufficient for our purpose that

$$\lim_{x\to\infty} p(x)\{y_n(x)y_m'(x) - y_m(x)y_n'(x)\} = 0, \tag{3.8.6}$$

and this certainly occurs if $p(x)$ has at most algebraic growth as $x \to \infty$, while all the eigenfunctions decay exponentially. This is the case with the Laguerre or Hermite equations.

It was pointed out in section 3.5 that at this stage our eigenfunctions contain an arbitrary multiplicative factor. We can specify this factor precisely by imposing the condition that for each n

$$\int_a^b r(x)\{y_n(x)\}^2 \, dx = 1 \tag{3.8.7}$$

and we then say that the set $\{y_n(x)\}$ is orthonormal. The functions sin nx and cos nx on $[-\pi, \pi]$ and the Legendre polynomials on $[-1, 1]$ give orthogonal sets which are not orthonormal. It will, however, be convenient in the subsequent analysis to assume that the functions have been so normalized.

If the coefficients $p(x)$, $q(x)$ and $r(x)$ and the constants which occur in the boundary conditions are all real, we may deduce that every eigenvalue is real. For if λ is a complex eigenvalue with eigenfunction $y_\lambda(x)$, it must then follow that $\bar{\lambda}$ is also an eigenvalue with eigenfunction $\overline{y_\lambda(x)}$. Thus (3.8.3) gives

$$(\lambda - \bar{\lambda}) \int_a^b r(x) \mid y_\lambda(x) \mid^2 dx = 0, \tag{3.8.8}$$

and since the integral is essentially positive,

$$\lambda = \bar{\lambda}, \tag{3.8.9}$$

which proves that λ is real.

3.9 Properties of approximation

In what follows we shall assume that $f(x)$ is any function for which

$$\int_a^b r(x)\{f(x)\}^2 \, dx \tag{3.9.1}$$

exists. We also assume that $\{y_n(x)\}$ is a given orthonormal set of functions. In general all integrals are to be taken in the sense of Lebesgue, but the reader can interpret them in the sense of Riemann.

The Schwartz inequality then gives

$$\left\{\int_a^b r(x)f(x)y_n(x)\,dx\right\}^2 \leqslant \left[\int_a^b r(x)\{f(x)\}^2\,dx\right]\left[\int_a^b r(x)\{y_n(x)\}^2\,dx\right]$$
$$= \int_a^b r(x)\{f(x)\}^2\,dx,$$

since by assumption the $\{y_n(x)\}$ are orthonormal. Thus for each n

$$a_n = \int_a^b r(x)f(x)y_n(x)\,dx \tag{3.9.2}$$

exists, and we shall designate this the nth 'Fourier coefficient' of $f(x)$.

Let $A_1, \ldots A_N$ be any set of constants and put

$$H(A_1, \ldots A_N) = \int_a^b r(x)\left\{f(x) - \sum_{k=1}^{N} A_k y_k(x)\right\}^2 dx. \tag{3.9.3}$$

If we then expand the right-hand side of (3.9.3) we obtain

$$\begin{aligned} H(A_1, \ldots A_N) &= \int_a^b r(x)\{f(x)\}^2\,dx - 2\sum_{k=1}^{N} A_k \int_a^b r(x)f(x)y_k(x)\,dx \\ &\qquad + \sum_{j=0}^{N}\sum_{k=0}^{N} A_jA_k \int_a^b r(x)y_j(x)y_k(x)\,dx, \\ &= \int_a^b r(x)\{f(x)\}^2\,dx - 2\sum_{k=1}^{N} A_k a_k + \sum_{k=1}^{N} A_k^2, \\ &= \int_a^b r(x)\{f(x)\}^2\,dx - \sum_{k=1}^{N} a_k^2 + \sum_{k=1}^{N} (a_k - A_k)^2. \end{aligned} \tag{3.9.4}$$

It therefore follows that out of all possible real values for the coefficients $A_1, A_2, \ldots A_N$, the choice $A_k = a_k$ for $k = 1, 2, \ldots N$ will minimize $H(A_1, A_2, \ldots A_N)$. With this choice

$$H(a_1, a_2, \ldots a_N) = \int_a^b r(x)\{f(x)\}^2\,dx - \sum_{k=1}^{N} a_k^2. \tag{3.9.5}$$

The left-hand side is non-negative and so we deduce the Bessel inequality

$$\int_a^b r(x)\{f(x)\}^2\,dx \geqslant \sum_{k=0}^{N} a_k^2. \tag{3.9.6}$$

In this inequality the left-hand side is independent of N, while the right-hand side is a monotone increasing sequence. It must follow therefore that

$$\lim_{N\to\infty} \sum_{k=0}^{N} a_k^2$$

exists, and the series converges.

If

$$\lim_{N\to\infty} \sum_{k=0}^{N} a_k^2 = \int_a^b r(x)\{f(x)\}^2\,dx,$$

so that $H \to 0$ as $N \to \infty$, then we say that

$$\sum_{k=0}^{\infty} a_k y_k(x)$$

converges in mean square to $f(x)$. If $\sum_{k=0}^{\infty} a_k y_k(x)$ converges uniformly to $f(x)$ on $[a, b]$, then obviously the integrand in (3.9.3) vanishes, and so H also converges in mean square to $f(x)$. The more interesting problem is when the convergence in mean square implies the convergence in some sense of $\sum_{k=0}^{\infty} a_k y_k(x)$ to $f(x)$, and this is examined in [15].

If, for every function $f(x)$ for which

$$\int_a^b r(x)\{f(x)\}^2 \, dx < \infty, \tag{3.9.7}$$

it is also equal to $\sum_{k=0}^{\infty} a_k^2$, we say that the orthonormal set $\{y_k(x)\}$ is complete or closed. This is equivalent to the statement that if $g(x)$ is a function of the class defined by (3.9.7) and if for every n

$$\int_a^b r(x)g(x)y_n(x) \, dx = 0, \tag{3.9.8}$$

then $g(x)$ vanishes (almost everywhere) on $[a, b]$.

If $\sum_{n=0}^{\infty} A_n y_n(x)$ converges uniformly to $f(x)$ on $[a, b]$, then the coefficients A_n are the Fourier coefficients a_n, for

$$a_m = \int_a^b r(x)f(x)y_m(x) \, dx = \sum_{n=0}^{\infty} A_n \int_a^b r(x)y_n(x)y_m(x) \, dx = A_m. \tag{3.9.9}$$

However, this result is not in general very useful as we can seldom assume uniform convergence.

3.10 The problem of completeness

The mathematical physicist requires that all well-behaved solutions of the more important partial differential equations can be expressed as sums or integrals of the functions that arise by separating the variables. It is shown in Appendices I and II that separation of variables leads to some of the Sturm–Liouville equations to be considered. Thus the problem of the completeness of the orthonormal sets which arise as eigenfunctions is of the first importance.

The regular case is comparatively easy. It may be shown (see [15],

Chapter I), that in this case the problem is equivalent to proving the completeness of the sets sin nx and cos nx, and this is the classical Fourier problem (see [18]).

The non-regular case is by far the most common and is very much more difficult. This is hardly surprising, since in general the representation will include both series of eigenfunctions and integrals over the continuous part of the spectrum. For a general analysis of the completeness of Sturm–Liouville functions the reader is referred to Chapters II and IX of [15].

In this volume we have considered only four such functions, and some results on their completeness may be obtained more simply.

The eigenfunctions of the Legendre equation are polynomials. Thus any arbitrary polynomial can be expressed as a finite sum of Legendre polynomials (see Chapter 8).

Now by the Weierstrass theorem [6] any function continuous on a closed interval can be uniformly approximated as closely as we choose by a polynomial. Hence given such a function and $\varepsilon > 0$ we can find an N, so that for some polynomial of degree N, say $f_N(x)$,

$$|f(x) - f_N(x)| < \varepsilon$$

for each $x \in [-1, 1]$. Since

$$f_N(x) = \sum_{n=0}^{N} a_n P_n(x)$$

it follows that

$$\int_{-1}^{1} \left| f(x) - \sum_{n=0}^{N} a_n P_n(x) \right|^2 dx \leqslant 2\varepsilon^2.$$

This establishes the completeness.

The completeness of the Laguerre and Hermite functions can be established in a somewhat similar way. In the case of the Laguerre polynomials we consider functions for which

$$\int_0^\infty e^{-x}\{f(x)\}^2 \, dx < \infty.$$

Then given $\delta > 0$ we can choose an X so large that

$$\int_X^\infty e^{-x}\{f(x)\}^2 \, dx < \delta.$$

The interval $[0, X]$ is finite, and so on this we can use the Weierstrass approximation theorem as above, and again any polynomial is the sum of Laguerre polynomials. A similar treatment can be applied to the Hermite functions.

These methods cannot be applied to the Bessel functions, since they are not polynomials. If we considered the most common problem of the representation of a function of suitable class on $[0, b]$, it turns out that this problem is 'almost regular'. Explicitly, we may apply the theorem for the regular case to the representation on $[a, b]$ for every $a > 0$. This representation has a uniform limit as $a \to +0$, which is itself a series in the correct eigenfunctions. It therefore follows that

$$\lim_{a\to 0} \int_a^b x\left\{f(x) - \sum_{n=0}^{\infty} a_n y_n(a, x)\right\}^2 \mathrm{d}x = 0,$$

and the eigenfunctions form a complete set.

Exercises

1. Find the adjoint of the following differential equations:
   (a) $y'' + 5y' - 6y = 0$;
   (b) $x^2y'' + (x + 3)y' + (5 - 4x)y = 0$;
   (c) $x(1 - x)y'' + \{c - (a + b + 1)x\}y' - aby = 0$.
2. Show that if $M(u)$ is the adjoint differential operator to

$$L(y) = a_0(x)y'' + a_1(x)y' + a_2(x)y,$$

then

$$u(x)L\{y(x)\} - y(x)M\{u(x)\} = \frac{\mathrm{d}}{\mathrm{d}x}\Big[a_0(x)u(x)y'(x) - y(x)\frac{\mathrm{d}}{\mathrm{d}x}\{a_0(x)u(x)\} + a_1(x)u(x)y(x)\Big].$$

Hence deduce that if $u_1(x)$ and $u_2(x)$ are independent solutions of $M(u) = 0$, then a general solution of the equation

$$L\{y(x)\} = f(x)$$

can be found by solving the pair of algebraic equations

$$a_0(x)u_1(x)y'(x) + [\{a_1(x) - a_0'(x)\}u_1(x) - a_0(x)u_1'(x)]y(x) = A_1 + \int_a^x u_1(\xi)f(\xi)\,\mathrm{d}\xi$$

$$a_0(x)u_2(x)y'(x) + [\{a_1(x) - a_0'(x)\}u_2(x) - a_0(x)u_2'(x)]y(x) = A_2 + \int_a^x u_2(\xi)f(\xi)\,\mathrm{d}\xi$$

for y, where A_1 and A_2 are constants, and a is a convenient lower limit of integration.

3. Suppose that $u(x, \xi)$ is a solution of the symmetric equation

$$\frac{\mathrm{d}}{\mathrm{d}x}\left\{p(x)\frac{\mathrm{d}y}{\mathrm{d}x}\right\} + q(x)u = 0,$$

where $u(x, \xi)$ satisfies the boundary conditions $u'(a, \xi) + hu(a, \xi) = 0$, $u'(b, \xi) + ku(b, \xi) = 0$, is continuous at $x = \xi$, and

$$\lim_{\varepsilon \to 0} \{u'(x + \varepsilon, \xi) - u'(x - \varepsilon, \xi)\} = -\frac{1}{p(\xi)}.$$

Deduce that u is a symmetric function of x and ξ.

Hence show that the solution of

$$\frac{\mathrm{d}}{\mathrm{d}x}\left\{p(x)\frac{\mathrm{d}y}{\mathrm{d}x}\right\} + q(x)y(x) = f(x)$$

which satisfies at a and b the same boundary conditions as $u(x, \xi)$ is

$$y(x) = \int_a^b u(x, \xi)f(\xi)\,\mathrm{d}\xi.$$

4. Transform the Bessel equation $x^2y'' + xy' + (x^2 - \nu^2)y = 0$, by the substitution $x = \mathrm{e}^{\xi}$. Hence deduce that any solution has an infinity of real zeros, whose separation approaches π as $x \to \infty$.

5. Transform the Legendre equation $(1 - x^2)y'' - 2xy' + \lambda^2 y = 0$ by the substitution $x = \tanh \xi$. Hence deduce that if $\lambda > 0$, and x_1, x_2 are a pair of consecutive zeros of any solution lying on the open interval (0, 1) with $x_2 > x_1$ then

$$x_2 - x_1 \geqslant (1 - x_2^{\,2}) \tanh (\pi/\lambda).$$

6. Find all the eigenvalues of the Sturm–Liouville system

$$4\frac{\mathrm{d}}{\mathrm{d}x}\left\{\mathrm{e}^{-x}\frac{\mathrm{d}y}{\mathrm{d}x}\right\} + (1 + \lambda)\mathrm{e}^{-x}y = 0,$$

$y(0) = 0$ when (a) $y(1) = 0$, (b) $y'(1) + y(1) = 0$.

7. Show that the eigenvalues of the Sturm–Liouville system

$$x\frac{\mathrm{d}}{\mathrm{d}x}\left\{x\frac{\mathrm{d}y}{\mathrm{d}x}\right\} + \lambda y = 0,$$

$y(a) = 0, \quad y(b) = 0, \quad b > a > 0,$
are given by $\lambda = n^2\pi^2/\{\log (b/a)\}^2$.

8. Show that if $\{y_n(x)\}$ are an orthonormal set on the interval $[a, b]$ with weighting function $r(x) > 0$, and $f(x)$ and $g(x)$ both permit expansions on the interval in a series of the eigenfunctions, and their Fourier coefficients are the sets $\{a_n\}$, $\{b_n\}$ respectively, then

$$\int_a^b r(x)f(x)g(x)\,\mathrm{d}x = \sum_{n=0}^{\infty} a_n b_n.$$

Answers

1. (a) $u'' - 5u' - 6u = 0$. (b) $x^2u'' + 3(x - 1)u' + 2(3 - 2x)u = 0$.
   (c) $x(1 - x)u'' + \{(a + b - 3)x + 2 - c\}u' - (a - 1)(b - 1)u = 0$.

6. (a) $\lambda = n^2\pi^2$. (b) $\lambda = \mu_n^{\,2}$, where $\tan \mu_n + 2\mu_n = 0$.

CHAPTER 4

Asymptotics

4.1 Asymptotic equivalence

If $f(x)$ and $g(x)$ are a pair of functions of the real variable x which satisfy

$$\lim_{x \to +\infty(-\infty)} f(x)/g(x) = 1,$$

we say that the functions are asymptotically equal or equivalent. Alternatively, we say that each function is asymptotic to the other as $x \to +\infty(-\infty)$. The relation is expressed symbolically by $f(x) \sim g(x)$. For example, $e^{1/x} \sim 1$, and $2 \cosh x \sim e^x$ as $x \to +\infty$, whilst $2 \cosh x \sim e^{-x}$ as $x \to -\infty$.

Before extending this concept to functions of the complex variable z, it is instructive to examine the function $2 \cosh z = e^z + e^{-z}$. Since $| e^z | = \exp\{\mathscr{R}(z)\}$, it follows that $2 \cosh z$ behaves like e^z or e^{-z} according as $| \arg z | < \pi/2$ or $\pi/2 < | \arg z | \leqslant \pi$. That is, a function may have different asymptotic representations in different sectors of the z-plane. In this example the two sectors are separated by the straight line $\arg z = \pm\pi/2$, and on this line $| e^z | = | e^{-z} |$. This behaviour, known as the Stokes phenomenon, is a general occurrence with asymptotic forms.

Formally, two functions $f(z)$ and $g(z)$ are asymptotically equal as $| z | \to \infty$ for every constant argument within the range $\theta_1 < \arg z < \theta_2$ if

$$\lim_{|z| \to \infty} f(z)/g(z) = 1. \tag{4.1.1}$$

This condition can be expressed in the equivalent form

$$\lim_{|z| \to \infty} \left| \frac{f(z) - g(z)}{g(z)} \right| = 0. \tag{4.1.2}$$

In this case we write

$$f(z) \sim g(z) \qquad \text{for } \theta_1 < \arg z < \theta_2. \tag{4.1.3}$$

The elementary transformation

$$\zeta = (z - a)^{-1}$$

maps the z-plane (1–1) on the ζ-plane in such a way that $\zeta \to \infty$ as $z \to a$ and the sector $-\theta_2 < \arg \zeta < -\theta_1$ corresponds to the sector $\theta_1 < \arg (z - a) < \theta_2$. Accordingly, it is possible to define asymptotic equivalence as $z \to a$, where a is any arbitrary point of the z-plane. However, we shall only be concerned with asymptotic values as $|z| \to \infty$.

It follows from the definition that if

$$f(z) \sim g_1(z) \qquad \text{for } \theta_1 < \arg z < \theta_2$$

and

$$f(z) \sim g_2(z) \qquad \text{for } \theta_3 < \arg z < \theta_4,$$

where $\theta_3 < \theta_2$ and $g_1(z) \neq g_2(z)$, then

$$g_1(z) \sim g_2(z) \qquad \text{for } \theta_3 < \arg z < \theta_2.$$

Returning to the previous example in which $2 \cosh z \sim e^z$ for $|\arg z| < \pi/2$, we note that $2 \cosh z \sim e^z + p_n(z)$ when $|\arg z| < \pi/2$, where $p_n(z)$ is any polynomial in z. That is, asymptotic equivalence is not a unique relation.

4.2 Asymptotic expansions

The ideas of the first section were generalized by Poincaré. We now suppose that we have a given function $f(z)$ and a sequence of functions $\{\phi_n(z)\}$. Then we say that $f(z)$ has an asymptotic expansion in the sense of Poincaré, which we symbolize by

$$f(z) \sim \sum_{n=0}^{\infty} a_n \phi_n(z) \qquad \theta_1 < \arg z < \theta_2 \tag{4.2.1}$$

if in the sector $\theta_1 < \arg z < \theta_2$ and for every N

$$\lim_{|z| \to \infty} \left| \frac{f(z) - \sum_{n=0}^{N} a_n \phi_n(z)}{\phi_N(z)} \right| \to 0. \tag{4.2.2}$$

An equivalent definition is that for every N and for $\theta_1 < \arg z < \theta_2$

$$f(z) = \sum_{n=0}^{N-1} a_n \phi_n(z) + O\{|\phi_N(z)|\} \tag{4.2.3}$$

as $|z| \to \infty$.

There is no assumption of convergence in these definitions, and in general the expansion will be strictly divergent.

As an example, consider the function $E_\nu(z)$, $\nu > 0$, which in $|\arg z| < \pi/2$ can be defined as

$$E_\nu(z) = \int_1^\infty \frac{e^{-zt}}{t^\nu}\, dt. \tag{4.2.4}$$

If we integrate once by parts we derive a recurrence formula

$$E_\nu(z) = -\frac{1}{z}\frac{e^{-zt}}{t^\nu}\bigg|_1^\infty - \frac{\nu}{z}\int_1^\infty \frac{e^{-zt}}{t^{\nu+1}}\, dt,$$

$$= z^{-1}\{e^{-z} - \nu E_{\nu+1}(z)\}. \tag{4.2.5}$$

If we apply this N times we obtain

$$E_\nu(z) = z^{-1}e^{-z}\left\{1 - \frac{\nu}{z} + \frac{\nu(\nu+1)}{z^2} - \dots \right.$$
$$\left. + (-1)^{N-1}\frac{\nu(\nu+1)\dots(\nu+N-2)}{z^{N-1}}\right\}$$
$$+ (-1)^N\frac{\nu(\nu+1)\dots(\nu+N-1)}{z^N}E_{\nu+N}(z). \tag{4.2.6}$$

Let us take $\phi_n(z) = z^{-(n+1)}e^{-z}$. Then comparing (4.2.6) with (4.2.3), we can put

$$E_\nu(z) \sim z^{-1}e^{-z}\sum_{n=0}^{\infty}(-1)^n\frac{\nu(\nu+1)\dots(\nu+n-2)}{z^n} \tag{4.2.7}$$

if we show that as $|z| \to \infty$

$$|z|^{-N}|E_{\nu+N}(z)| \quad \text{is} \quad O|\phi_N(z)|.$$

If, however, $|\arg z| \leqslant \pi/2 - \delta$,
we have

$$|E_{\nu+N}(z)| \leqslant \int_1^\infty \frac{e^{-xt}}{t^{\nu+N}}dt \leqslant \int_1^\infty e^{-xt}\,dt = \frac{e^{-x}}{x} \leqslant \frac{|e^{-z}|}{|z|}\operatorname{cosec}\delta,$$

and so the equivalence (4.2.7) is established.

This example is quite instructive, as it shows the use of an asymptotic series. In this case

$$|a_{n+1}\phi_{n+1}(z)|/|a_n\phi_n(z)| = (\nu+n)/|z|. \tag{4.2.8}$$

Every individual term in the expansion (4.2.7) tends to zero as $z \to \infty$ for $|\arg z| < \pi/2$. On the other hand, for a fixed z, the successive terms decrease in modulus until n reaches the value which satisfies the inequality $\nu + n - 1 < |z| < \nu + n$. After this stage the moduli of successive terms increase, and ultimately diverge to infinity. We know, however, by the definition of an asymptotic series that the error obtained by terminating the series at a definite term is of the order of the next term in the expansion. Thus if we arrange that this term, and therefore the order of the error, is the smallest term, we have the 'best possible' representation of the original function.

Suppose in our example we consider the case of z real and put $z = x$; then the error in terminating the series is

$$\frac{\nu(\nu+1)\ldots(\nu+N-1)}{x^N}E_{\nu+N}(x) \leqslant \frac{\nu(\nu+1)\ldots(\nu+N-1)}{x^{N+1}}e^{-x}. \tag{4.2.9}$$

The best representation will be obtained if we terminate the series when

$$\frac{\nu+N-1}{x} \leqslant 1 \quad \text{and} \quad \frac{\nu+N}{x} > 1.$$

Suppose, for example, we take $\nu = 1$; then we terminate the series at the Nth term, where $N \leqslant x < N+1$. For $x = N$, an estimate of the error is $(N!/N^{N+1})e^{-N}$ and applying Stirling's approximation for $N!$ (see section 5.9), we see that this estimate is approximately $(2\pi/N)^{1/2}e^{-2N}$. Accordingly if we use the asymptotic series to calculate $E_1(N)$, the result will always be in error, but we can arrange that the error has as an approximate upper bound the estimate given in Table 4.1.

TABLE 4.1

N	2	3	5	10	20
Estimate of errors	0·0325	$3{\cdot}6 \times 10^{-3}$	$5{\cdot}1 \times 10^{-5}$	$1{\cdot}6 \times 10^{-9}$	$2{\cdot}4 \times 10^{-18}$
Per cent error	66·5	27·5	4·44	0·038	$2{\cdot}44 \times 10^{-8}$

We see therefore that there is an essential difference between a convergent series and an asymptotic series. In principle, by taking enough terms into the calculation, a function defined by a convergent series can be computed to any order of accuracy. However, a function defined by an asymptotic series has a residual error for any given value of the variable. It is not possible to avoid this, though in practice it might be very small indeed.

4.3 Properties of asymptotic power series

In the example in the last section, $\phi_n(z) = z^{-(n+1)}e^{-z}$, and so we may express the result in the form

$$ze^zE_\nu(z) \sim \sum_{n=0}^{\infty}\frac{a_n}{z^n} \quad \text{in} \quad |\arg z| < \pi/2. \tag{4.3.1}$$

The expression on the right-hand side is called an asymptotic power series, and we now consider some of the properties of such series. In what follows we shall assume that in some common sector

$$\theta_1 < \arg z < \theta_2$$

$$f(z) \sim \sum_{n=0}^{\infty} a_n z^{-n} \quad \text{and} \quad g(z) \sim \sum_{n=0}^{\infty} b_n z^{-n}.$$

(a) If k is a constant (in general complex), then

$$kf(z) \sim \sum_{n=0}^{\infty} ka_n z^{-n}. \tag{4.3.2}$$

This is obvious, for if $\left| f(z) - \sum_{n=0}^{N-1} a_n z^{-n} \right|$ is $O\{|z|^{-N}\}$, so is $\left| kf(z) - \sum_{n=0}^{N-1} ka_n z^{-n} \right|$.

(b)
$$f(z) + g(z) \sim \sum_{n=0}^{\infty} (a_n + b_n) z^{-n}. \tag{4.3.3}$$

This again is immediate, since

$$\left| f(z) + g(z) - \sum_{n=0}^{N-1} (a_n + b_n) z^{-n} \right| \leqslant \left| f(z) - \sum_{n=0}^{N-1} a_n z^{-n} \right| + \left| g(z) - \sum_{n=0}^{N-1} a_n z^{-n} \right| = O\{|z|^{-N}\}.$$

It is clear that both the properties (a) and (b) can be generalized to asymptotic series, which are not asymptotic power series, provided the sequence $\{\phi_n(z)\}$ is the same for both series. If $f(z) \sim \sum_{n=0}^{\infty} a_n \phi_n(z)$ and $g(z) \sim \sum_{n=0}^{\infty} b_n \psi_n(z)$ for $\theta_1 < \arg z < \theta_2$, we can often produce a composite set $\{\phi_n\} \cup \{\psi_n\}$ and regard each expansion as in the composite sequence, with zero coefficients.

The remaining properties apply only to asymptotic power series.

(c)
$$f(z)g(z) \sim \sum_{n=0}^{\infty} c_n z^{-n} \quad \text{where} \quad c_n = \sum_{k=0}^{n} a_{n-k} b_k. \tag{4.3.4}$$

Let

$$f(z) = \sum_{n=0}^{N-1} a_n z^{-n} + z^{-N} R_N(z) \quad \text{and} \quad g(z) = \sum_{n=0}^{N-1} b_n z^{-n} + z^{-N} S_N(z), \tag{4.3.5}$$

so that our assumption is equivalent to the statement that for $\theta_1 < \arg z < \theta_2$ and for $|z|$ greater than some constant r_0

$$|R_N(z)| < K_N \quad \text{and} \quad |S_N(z)| < K_N,$$

where K_N is independent of z, but will in general depend on N.

Then

$$f(z)g(z) = \sum_{n=0}^{N-1} c_n z^{-n} + z^{-N} T_N(z),$$

where

$$T_N(z) = \sum_{n=1}^{N-1} \sum_{m=N-n+1}^{N-1} a_n b_m z^{-N-n-m} + S_N(z) \sum_{n=0}^{N-1} a_n z^{-n} + R_N \sum_{n=0}^{N-1} b_n z^{-n} + z^{-N} R_N(z) S_N(z),$$

and so re-summing the first term by diagonals gives us

$$|T_N(z)| \leqslant \sum_{r=1}^{N-1} \sum_{n=r}^{N-1} |a_n|\,|b_{N+r-n-1}|\, r_0^{1-r} + K_N \sum_{n=0}^{N-1} \{|a_n| + |b_n|\} r_0^{-n} + K_N^2 r_0^{-N}.$$

Thus $|T_N(z)|$ is bounded and so (4.3.4) is established.

(d) If $a_0 = a_1 = \ldots = a_{m-1} = 0$ and $a_m \neq 0$, then

$$\{f(z)\}^{-1} \sim \frac{z^m}{a_m} \sum_{n=0}^{\infty} \frac{d_n}{z^n} \tag{4.3.6}$$

where $d_0 = 1$ and $\sum_{k=0}^{n} d_{n-k} a_{m+k} = 0$.

As $|z| \to \infty$, $z^m f(z) \to a_m$ and hence

$$z^{-m}\{f(z)\}^{-1} \to a_m^{-1}.$$

It follows that

$$z^{-m}\{f(z)\}^{-1} \sim a_m^{-1} \sum_{n=0}^{\infty} \frac{d_n}{z^n}$$

where $d_0 = 1$. The application of the multiplicative properties then shows that

$$a_m z^{-m} = \left(\sum_{n=m}^{\infty} a_n z^{-n}\right)\left(\sum_{n=0}^{\infty} d_n z^{-n}\right) = z^{-m}\left(\sum_{n=0}^{\infty} a_{n+m} z^{-n}\right)\left(\sum_{n=0}^{\infty} d_n z^{-n}\right)$$

and hence if $n \geqslant 1$

$$\sum_{k=0}^{n} d_{n-k} a_{m+k} = 0.$$

Thus the coefficients can be calculated recursively.

(e)

$$\int_z^\infty \{f(\zeta) - a_0 - a_1/\zeta\}\,\mathrm{d}\zeta \sim \sum_{n=2}^{\infty} \frac{a_n}{(n-1)z^{n-1}}. \tag{4.3.7}$$

By assumption, for $\theta_1 < \arg z < \theta_2$ and for $|z|$ greater than some constant r_0,

$$\left| f(z) - \sum_{n=0}^{N-1} a_n z^{-n} \right| \leqslant \frac{K_N}{|z|^N}$$

where again K_N is independent of z. Then if we join z to infinity by a contour which is contained entirely within the above region,

$$\left| \int_z^\infty \left\{ f(\zeta) - \sum_{n=0}^{N-1} a_n \zeta^{-n} \right\} \mathrm{d}\zeta \right| \leqslant K_N \int_{|z|}^\infty \frac{\mathrm{d}\zeta}{\zeta^N} = \frac{K_N}{(N-1)\,|z|^{N-1}}.$$

It follows that

$$\left| \int_z^\infty \{f(\zeta) - a_0 - a_1/\zeta\}\,\mathrm{d}\zeta - \sum_{n=2}^{N-1} \frac{a_n}{(n-1)z^{n-1}} \right| \leqslant \frac{K_N}{(N-1)\,|z|^{N-1}}$$

and (4.3.7) can be deduced.

(f) If $f(z)$ is regular in $|z| > r_0$ and for $\theta_1 < \arg z < \theta_2$, then

$$f'(z) \sim -\sum_{n=1}^{\infty} \frac{na_n}{z^{n+1}} \tag{4.3.8}$$

in every interior sector $\theta_1 + \delta \leqslant \arg z \leqslant \theta_2 - \delta$, for $\delta > 0$.

If once again we write

$$f(z) = \sum_{n=0}^{N-1} a_n z^{-n} + z^{-N} R_N(z),$$

then

$$f'(z) = -\sum_{n=1}^{N-1} na_n z^{-n-1} + z^{-N} S_N(z),$$

where

$$S_N(z) = R_N'(z) - \frac{N}{z} R_N(z).$$

We know that for $\theta_1 < \arg z < \theta_2$, $|R_N(z)| < K_N$. Let now z_0 be a point within $\theta_1 + \delta \leqslant \arg z_0 \leqslant \theta_2 - \delta$, $|z_0| \geqslant r_0$. Then if we choose $\varepsilon < \sin \delta/2$, every point of the circle C, $|z - z_0| = \varepsilon\,|z_0|$, lies within the wider sector $\theta_1 + \delta/2 < \arg z < \theta_2 - \delta/2$.

Then the Cauchy integral formula

$$R_N'(z_0) = \frac{1}{2\pi i}\int_C \frac{R_N(z)}{(z - z_0)^2}\, dz$$

implies the inequality

$$|\, R_N'(z) \,| \leqslant \frac{K_N}{\varepsilon \,|\, z_0 \,|}.$$

Hence $|\, S_N(z_0) \,| \leqslant K_N r_0^{-1}\{N + \varepsilon^{-1}\}$, as required.

(g) If $f(z)$ is regular for all $|\, z \,| \geqslant r_0$ and $f(z) \sim \sum_{n=0}^{\infty} a_n/z^n$ for all arg z, then the asymptotic series converges and is an ordinary Laurent series.

Since $f(z)$ is regular for $|\, z \,| \geqslant r_0$ it has a Laurent expansion

$$\sum_{n=-\infty}^{\infty} A_n z^{-n}, \tag{4.3.9}$$

where

$$2\pi i A_n = \int_{|\zeta| = r_1} f(\zeta)\zeta^{n-1}\, d\zeta \quad \text{and} \quad r_1 > r_0.$$

Now as $|\, z \,| \to \infty$, $f(z) \to a_0$, from the asymptotic series, and so for some r_2 if

$$r > r_2, \qquad |f(z)\,| < M,$$

where M is some constant. Thus if $r_1 > r_2$,

$$|\, A_n \,| \leqslant 2 \,|\, a_0 \,|\, r_1^n,$$

and since r_1 can be as large as we like,

$$A_n = 0 \qquad n = -1, -2, \ldots.$$

Then on letting $|\, z \,| \to \infty$, $A_0 = a_0$, and similarly

$$a_1 = \lim_{|z| \to \infty} |\, z \,| \{f(z) - a_0\} = A_1,$$

and so on successively.

(h) An asymptotic power series is unique. That is, if

$$f(z) \sim \sum_{n=0}^{\infty} a_n z^{-n} \text{ and } f(z) \sim \sum_{n=0}^{\infty} b_n z^{-n} \quad \text{for} \quad \theta_1 < \arg z < \theta_2,$$

then $a_n = b_n$ for every n.

Within the sector

$$a_0 = \lim_{|z| \to \infty} f(z) = b_0$$

and then

$$a_1 = \lim_{|z| \to \infty} z\{f(z) - a_0\} = b_1,$$

and so on.

4.4 Watson's lemma

Suppose $f(t)$ is regular in the neighbourhood of $t = 0$, so that it has an expansion for small t of the form

$$f(t) = \sum_{s=0}^{\infty} a_s t^s. \tag{4.4.1}$$

Suppose further that corresponding to a given real number $\nu > 0$, we can find a pair of constants K and r_0, such that for all t in the range $0 \leqslant t < \infty$

$$|f(t)| \leqslant K \exp(r_0 t^\nu).$$

Then as $|z| \to \infty$ within the sector $|\arg z| \leqslant \pi/2 - \delta$,

$$\nu \int_0^\infty \exp(-zt^\nu) f(t)\,\mathrm{d}t \sim \sum_{s=0}^{\infty} \Gamma\left(\frac{s+1}{\nu}\right) a_s z^{-(1+s)/\nu}. \tag{4.4.2}$$

By our initial assumptions

$$F_N(t) = \left| f(t) - \sum_{s=0}^{N-1} a_s t^s \right| \leqslant K_N t^N \exp(r_0 t^\nu),$$

where K_N is some new constant that will in general depend on N. Thus

$$\left| \int_0^\infty \exp(-zt^\nu) f(t)\,\mathrm{d}t - \sum_{s=0}^{N-1} a_s \int_0^\infty t^s \exp(-zt^\nu)\,\mathrm{d}t \right| \leqslant |z|^{-(N+1)/\nu} |R_N(z)|,$$

where

$$|R_N(z)| = |z|^{(N+1)/\nu} \int_0^\infty \exp(-xt^\nu) F_N(t)\,\mathrm{d}t$$

$$\leqslant K_N |z|^{(N+1)/\nu} \int_0^\infty t^N \exp\{-(x - r_0)t^\nu\}\,\mathrm{d}t.$$

However

$$\nu \int_0^\infty t^s \exp(-\lambda t^\nu)\,\mathrm{d}t = \lambda^{-(s+1)/\nu} \int_0^\infty \mathrm{e}^{-u} u^{-1+(s+1)/\nu}\,\mathrm{d}u$$

$$= \lambda^{-(s+1)/\nu}\,\Gamma\left(\frac{s+1}{\nu}\right).$$

Thus

$$|R_N(z)| \leqslant K_N\,\Gamma\left(\frac{N+1}{\nu}\right)\{|z|/(x - r_0)\}^{(s+1)/\nu}$$

$$\leqslant K_N\,\Gamma\left(\frac{N+1}{\nu}\right)(\operatorname{cosec}\delta)^{(s+1)/\nu},$$

and so

$$\nu \int_0^\infty \exp(-zt^\nu) f(t)\, dt \sim \sum_{s=0}^{\infty} a_s \nu \int_0^\infty t^s \exp(-zt^\nu)\, dt$$

$$= \sum_{s=0}^{\infty} a_s \Gamma\left(\frac{s+1}{\nu}\right) z^{-(s+1)/\nu}.$$

4.5 Irregular points of ordinary differential equations

It was shown in Chapter 1 that in the neighbourhood of either an ordinary point or a regular singular point, the solutions of an ordinary differential equation can be represented by convergent expansions. A simple example will show that this is not in general true for an irregular point.

The equation

$$z^3 \frac{d^2 w}{dz^2} + z(z+1)\frac{dw}{dz} - w = 0 \tag{4.5.1}$$

has an irregular singular point at $z = 0$. It may easily be verified that a solution of this equation is

$$\exp(1/z) E_1(1/z), \tag{4.5.2}$$

where $E_1(\zeta)$ is defined by equation (4.2.4). We showed in section 4.2 that $e^\zeta E_1(\zeta)$ had an asymptotic expansion for large $|\zeta|$, but not a convergent expansion. The same is true, therefore, for this particular solution of equation (4.5.1) near the origin.

We consider the differential equation

$$w'' + p(z)w' + q(z)w = 0 \tag{4.5.3}$$

and suppose that it has an irregular point at infinity. This case occurs frequently with the differential equations of mathematical physics. We shall further assume that for large $|z|$, either $p(z)$ and $q(z)$ have Laurent expansions

$$p(z) = \sum_{s=0}^{\infty} p_s z^{-s} \tag{4.5.4}$$

and

$$q(z) = \sum_{s=0}^{\infty} q_s z^{-s} \tag{4.5.5}$$

or they have asymptotic power series of the form (4.5.4) and (4.5.5), in some sector of the z-plane. We therefore exclude equations in which $p(z)$ and $q(z)$ are unbounded for large $|z|$.

We substitute formally into (4.5.3) the asymptotic expansion

$$w(z) \sim e^{\lambda z} z^{-\alpha} \sum_{s=0}^{\infty} c_s z^{-s}. \tag{4.5.6}$$

Since (4.5.6) is valid within some sector we can differentiate term by term, in any interior region. Thus we derive

$$w'(z) \sim e^{\lambda z} z^{-\alpha} \sum_{s=0}^{\infty} \{\lambda c_s - (s + \alpha - 1)c_{s-1}\} z^{-s} \tag{4.5.7}$$

and

$$w''(z) \sim e^{\lambda z} z^{-\alpha} \sum_{s=0}^{\infty} \{\lambda^2 c_s - 2\lambda(s + \alpha - 1)c_{s-1} + (s + \alpha - 2)(s + \alpha - 1)c_{s-2}\} z^{-s}, \tag{4.5.8}$$

where for convenience we write $c_s = 0$ for $s < 0$. We substitute (4.5.4–4.5.8) into (4.5.3) and make use of the results for the multiplication and addition of asymptotic series. On simplification we obtain an asymptotic expansion of the form

$$e^{\lambda z} z^{-\alpha} \sum_{s=0}^{\infty} A_s z^{-s} \sim 0, \tag{4.5.9}$$

where

$$A_0 = (\lambda^2 + p_0\lambda + q_0)c_0, \tag{4.5.10}$$

$$A_1 = (\lambda^2 + p_0\lambda + q_0)c_1 + \{\lambda p_1 + q_1 - (2\lambda + p_0)\alpha\}c_0, \tag{4.5.11}$$

and for $s \geqslant 2$

$$\begin{aligned} A_s = (\lambda^2 + p_0\lambda + q_0)c_s &+ \{\lambda p_1 + q_1 - (2\lambda + p_0)(s - 1 + \alpha)\}c_{s-1} \\ &+ \{\lambda p_2 + q_2 + (s - 2 + \alpha)(s - 1 + \alpha - p_1)\}c_{s-2} \\ &+ \sum_{k=3}^{s} \{\lambda p_k + q_k - (s - k + \alpha)p_{k-1}\}c_{s-k}. \end{aligned} \tag{4.5.12}$$

If we can choose the constants λ and α and the coefficients such that $A_s = 0$ for every s, then we have a formal asymptotic expansion for the solution. Since we have established properties of addition, multiplication and differentiation, we are able to assert that if the solution has an asymptotic series of this form, the process outlined above must derive it. We have not proved that the equation has got a solution which has an asymptotic expansion of this form.

4.6 The case of unequal roots

We see from (4.5.10) that $A_0 = 0$ only if λ is a root of

$$\lambda^2 + p_0\lambda + q_0 = 0. \tag{4.6.1}$$

We shall assume in this section that (4.6.1) has two unequal roots λ_1 and λ_2. It is clear that we may use either root in the subsequent analysis, and so formally generate two independent solutions.

When we have fixed λ, (4.5.11) shows that $A_1 = 0$ only if

$$\alpha = (\lambda p_1 + q_1)/(2\lambda + p_0) \tag{4.6.2}$$

and so the process will fail if $p_0 = -2\lambda$, unless also $\lambda p_1 + q_1 = 0$. If, however, $p_0 = -2\lambda_1$, the roots of (4.6.1) are equal. Then (4.6.1) is a perfect square, and this is the case we have temporarily excluded.

The condition that $A_2 = 0$ now gives, from (4.5.12), a relation between c_1 and c_0, namely

$$(2\lambda + p_0)c_1 = \{\lambda p_2 + q_2 + \alpha(\alpha + 1 - p_1)\}c_0. \tag{4.6.3}$$

The coefficients of the other terms can then be calculated from the recurrence relation with $s \geqslant 3$:

$$(2\lambda + p_0)(s-1)c_{s-1} = \{\lambda p_2 + q_2 + (s-2+\alpha)(s-1+\alpha-p_1)\}c_{s-2} + \sum_{k=3}^{s} \{\lambda p_k + q_k - (s-k+\alpha)p_{k-1}\}c_{s-k}. \tag{4.6.4}$$

For a full discussion of asymptotic solutions of differential equations see [16].

4.7 Application to the Whittaker and Bessel equations

For the notation of this section the reader is referred to Chapters 7 and 9, on the confluent hypergeometric functions and on the Bessel function.

Consider the Whittaker equation

$$w'' - \left\{\tfrac{1}{4} - \frac{k}{z} - \frac{1-4m^2}{4z^2}\right\}w = 0. \tag{4.7.1}$$

The quadratic corresponding to (4.6.1) is

$$\lambda^2 - \tfrac{1}{4} = 0. \tag{4.7.2}$$

If we take the root $\lambda = -\frac{1}{2}$, then (4.6.2) gives

$$\alpha = -k. \tag{4.7.3}$$

The recurrence relation between the coefficients for $s \geqslant 2$ is

$$(s-1)c_{s-1} = \{m^2 - \tfrac{1}{4} - (s-k-2)(s-k-1)\}c_{s-2} \tag{4.7.4}$$

or

$$4(s-1)c_{s-1} = \{4m^2 - (2s-2k-3)^2\}c_{s-2}. \tag{4.7.5}$$

This will give the asymptotic expansion

$$w(z) \sim e^{-z/2}z^k\left[1 + \frac{4m^2 - (2k-1)^2}{1!\,(4z)} + \frac{\{4m^2-(2k-1)^2\}\{4m^2-(2k-3)^2\}}{2!\,(4z)^2} + \dots\right] \tag{4.7.6}$$

and this is shown in Chapter 7 to be the asymptotic expansion of the Whittaker function $W_{k,m}(z)$.

We may similarly derive a solution with $\lambda = \frac{1}{2}$, $\alpha = k$, and this gives the asymptotic form of the second solution $W_{-k,m}(-z)$.

A second example is provided by the Bessel equation

$$w'' + z^{-1}w' + (1 - \nu^2 z^{-2})w = 0. \tag{4.7.7}$$

Here the quadratic is

$$\lambda^2 + 1 = 0. \tag{4.7.8}$$

If we take the root $\lambda = \mathrm{i}$, we find that

$$\alpha = \tfrac{1}{2}. \tag{4.7.9}$$

The recurrence relations reduce to

$$2\mathrm{i}(s-1)c_{s-1} = \{(s-3/2)^2 - \nu^2\}c_{s-2} \tag{4.7.10}$$

and hence there is a solution with the asymptotic form

$$\mathrm{e}^{\mathrm{i}z}z^{-1/2}\left\{1 + \frac{(1-4\nu^2)}{1!\,(8\mathrm{i}z)} + \frac{(1-4\nu^2)(9-4\nu^2)}{2!\,(8\mathrm{i}z)^2} + \dots\right\}. \tag{4.7.11}$$

This is the asymptotic expansion of

$$\left(\frac{\pi}{2}\right)^{1/2} \exp\left\{(\nu - \tfrac{1}{2})\frac{\pi\mathrm{i}}{2}\right\} H_\nu^{(1)}(z), \tag{4.7.12}$$

which is shown in Chapter 9 to be a solution of the equation. Similarly the roots $\lambda = -\mathrm{i}$, $\alpha = \frac{1}{2}$, would lead to the asymptotic expansion of

$$\left(\frac{\pi}{2}\right)^{1/2} \exp\left\{-(\nu - \tfrac{1}{2})\frac{\pi\mathrm{i}}{2}\right\} H_\nu^{(2)}(z). \tag{4.7.13}$$

4.8 The case of equal roots

If the quadratic (4.6.1) has a pair of equal roots λ_0 the transformation

$$w(z) = \mathrm{e}^{\lambda_0 z}v(z) \tag{4.8.1}$$

reduces the equation to one for $v(z)$, which has two zero roots for its corresponding quadratic. There is therefore no loss of generality if we assume $\lambda_0 = 0$, and hence $p_0 = q_0 = 0$.

If also $q_1 = 0$, then ∞ is a regular singular point, and this has been discussed in Chapter 1. We may therefore assume that $q_1 \neq 0$.

Under the transformation $z = \frac{1}{4}\zeta^2$ (4.5.3) becomes

$$\frac{\mathrm{d}^2 w}{\mathrm{d}\zeta^2} - \left\{\frac{\zeta}{2}p\left(\frac{\zeta^2}{4}\right) - \frac{1}{\zeta}\right\}\frac{\mathrm{d}w}{\mathrm{d}\zeta} + \frac{\zeta^2}{4}q\left(\frac{\zeta^2}{4}\right)w = 0. \tag{4.8.2}$$

This equation, with ζ as independent variable, is now of the same form as that discussed in section 4.5. If therefore we look for solutions

$$w(\zeta) = \mathrm{e}^{\lambda\zeta}\zeta^{-\alpha}\sum_{n=0}^{\infty} c_s\zeta^{-s}, \tag{4.8.3}$$

we derive first

$$\lambda^2 + q_1 = 0 \tag{4.8.4}$$

and then

$$\alpha = \tfrac{1}{2}(2p_1 - 1) \tag{4.8.5}$$

and the recurrence relations are simplified by the fact that the coefficient of $dw/d\zeta$ is an odd function of ζ^{-1}, and that of w is an even function. Since $q_1 \neq 0$, (4.8.4) must have distinct roots, and the theory then goes through as before.

For a fuller account of the theory see [16].

4.9 An illustration of the case of equal roots

If we take the equation

$$w'' + z^{-1}w = 0, \tag{4.9.1}$$

the auxiliary quadratic reduces to $\lambda^2 = 0$.

The transformation $z = \frac{1}{4}\zeta^2$ gives

$$\frac{d^2w}{d\zeta^2} - \frac{1}{\zeta}\frac{dw}{d\zeta} + w = 0. \tag{4.9.2}$$

We can solve this by the method previously outlined, but it is easy to see that if $w(\zeta) = \zeta v(\zeta)$, then

$$\frac{d^2v}{d\zeta^2} + \frac{1}{\zeta}\frac{dv}{d\zeta} + \left(1 - \frac{1}{\zeta^2}\right)v = 0 \tag{4.9.3}$$

and the solutions for $v(\zeta)$ are those of (4.7.12) and (4.7.13), with $\nu = 1$ and z replaced by ζ.

4.10 The W.K.B. method

Another problem of interest, particularly in quantum mechanics, is to find the asymptotic form of the solution of a Sturm–Liouville equation for a large value of the eigenvalue. A technique which does this was given originally in its simplest form by Jeffreys [8], but it is usually associated with the names of Wentzel [17], Kramers [9] and Brillouin [3].

We take the equation in the usual form

$$\frac{d}{dx}\left\{p(x)\frac{dy}{dx}\right\} + \{\lambda r(x) + q(x)\}y = 0. \tag{4.10.1}$$

Initially we investigate solutions in the neighbourhood of a point, which we can take as $x = 0$, and in this neighbourhood we assume that $p(x) > 0$ and $r(x)$ does not change sign. To make the analysis quite unambiguous we assume $r(x) > 0$, though the modifications required if it is negative are easily made. We are interested in the case when λ is very large compared with unity, and for convenience we shall put $\lambda = \mu^2$. The transformations

$$y(x) = \{r(x)p(x)\}^{-1/4}Y(\xi), \qquad \xi = \int_0^x \{r(t)/p(t)\}^{1/2}\,dt$$

reduce the equation to the normal form

$$\frac{d^2Y}{d\xi^2} + \{\mu^2 + Q(\xi)\}Y = 0, \tag{4.10.2}$$

where $Q(\xi)$ is a complicated expression involving $q(x)$ and the derivatives of $p(x)$ and $r(x)$, whose exact form is irrelevant to the argument.

It is to be noticed that the assumptions we have made about $p(x)$ and $r(x)$ ensure that the mapping of the x interval onto the ξ interval is (1–1) and the origins map into each other.

We rewrite (4.10.2) in the form

$$\frac{d^2Y}{d\xi^2} + \mu^2 Y = -Q(\xi)Y(\xi) \tag{4.10.3}$$

and so, if we solve this by the method of variation of parameters, a solution of (4.10.2) with the boundary conditions

$$Y(0) = \cos\alpha, \qquad Y'(0) = -\sin\alpha,$$

must satisfy the integral equation

$$Y(\xi) = \cos\alpha\cos\mu\xi - \frac{1}{\mu}\sin\alpha\sin\mu\xi - \frac{1}{\mu}\int_0^\xi \sin\mu(\xi-\eta)Q(\eta)Y(\eta)\,d\eta. \tag{4.10.4}$$

Thus, provided $Q(\xi)$ is bounded on the interval, $Y(\xi)$ will certainly be bounded on the interval. Then

$$Y(\xi) = \cos\alpha\cos\mu\xi + O\left(\frac{1}{\mu}\right). \tag{4.10.5}$$

If we now write

$$Y_1(\xi) = \mu\{Y(\xi) - \cos\alpha\cos\mu\xi\},$$

the integral equation becomes

$$Y_1(\xi) = -\sin\alpha\sin\mu\xi - \cos\alpha\int_0^\xi \sin\mu(\xi-\eta)Q(\eta)\cos\mu\eta\,d\eta - \frac{1}{\mu}\int_0^\xi \sin\mu(\xi-\eta)Q(\eta)Y_1(\eta)\,d\eta \tag{4.10.6}$$

and repetition of the same argument leads to the conclusion that $Y_1(\xi)$ is bounded and hence

$$Y_1(\xi) = -\sin\alpha\sin\mu\xi - \cos\alpha\int_0^\xi \sin\mu(\xi-\eta)Q(\eta)\cos\mu\eta\,\mathrm{d}\eta + O\left(\frac{1}{\mu}\right). \tag{4.10.7}$$

Accordingly

$$Y(\xi) = \cos\alpha\cos\mu\xi - \frac{1}{\mu}\left\{\sin\alpha\sin\mu\xi + \cos\alpha\int_0^\xi \sin(\xi-\eta)Q(\eta)\cos\mu\eta\,\mathrm{d}\eta\right\} + O\left(\frac{1}{\mu^2}\right), \tag{4.10.8}$$

and we may clearly continue the iteration.

4.11 The Legendre equation

This equation is

$$\frac{\mathrm{d}}{\mathrm{d}x}\left\{(1-x^2)\frac{\mathrm{d}y}{\mathrm{d}x}\right\} + \nu(\nu+1)y = 0. \tag{4.11.1}$$

The transformations of the previous section take the particular form

$$y(x) = (1-x^2)^{-1/4}Y(\xi) \quad\text{and}\quad x = \sin\xi, \tag{4.11.2}$$

so that the interval $(-1, 1)$ is mapped (1–1) onto $(-\pi/2, \pi/2)$ with the origins corresponding to each other. Equation (4.10.2) then is

$$\frac{\mathrm{d}^2Y}{\mathrm{d}\xi^2} + \{\nu(\nu+1) + \tfrac{1}{4}(1+\sec^2\xi)\}Y = 0 \tag{4.11.3}$$

which we can write as

$$\frac{\mathrm{d}^2Y}{\mathrm{d}\xi^2} + (\nu+\tfrac{1}{2})^2Y = -\tfrac{1}{4}\sec^2\xi\, Y. \tag{4.11.4}$$

We must assume that the right-hand side is bounded, and this restricts the solution to the interval $(-\pi/2+\delta, \pi/2-\delta)$. Then if we put $\mu = \nu + \frac{1}{2}$, the solution (4.10.8) gives in this case

$$Y(\xi) = \cos\alpha\cos\mu\xi - \frac{1}{\mu}\left\{\sin\alpha\sin\mu\xi - \frac{\cos\alpha}{4}\int_0^\xi \frac{\sin\mu(\xi-\eta)\cos\mu\eta}{\cos^2\eta}\,\mathrm{d}\eta\right\} + O\left(\frac{1}{\mu^2}\right). \tag{4.11.5}$$

Furthermore, since

$$2\sin\mu(\xi-\eta)\cos\mu\eta = \sin\mu\xi + \sin\mu\xi\cos 2\mu\eta - \cos\mu\xi\sin 2\mu\eta,$$

we have

$$\int_0^\xi \cos 2\mu\eta\sec^2\eta\,\mathrm{d}\eta = \frac{1}{2\mu}\sin 2\mu\xi\sec^2\xi - \frac{1}{\mu}\int_0^\xi \sin 2\eta\mu\sec^2\eta\tan\eta\,\mathrm{d}\eta$$
$$= O\left(\frac{1}{\mu}\right)$$

and similarly

$$\int_0^{\xi} \sin 2\mu\eta \sec^2 \eta \, d\eta = O\left(\frac{1}{\mu}\right) \quad \text{for } -\pi/2 + \delta \leqslant \xi \leqslant \pi/2 - \delta.$$

It follows that

$$Y(\xi) = \cos \alpha \cos \mu\xi - \frac{\sin \mu\xi}{\mu}\{\sin \alpha - \tfrac{1}{8} \cos \alpha \tan \xi\} + O\left(\frac{1}{\mu^2}\right). \quad (4.11.6)$$

In particular, if ν is an integer, one solution of the equation is the Legendre polynomial. If it is an even integer the polynomial is even, and so its derivative vanishes at $x = 0$. Thus the polynomial is a multiple of the solution above with $\alpha = \pi/2$, but the multiplying constant depends strongly on ν. In the case of an odd integer we simply have a multiple of the solution above with $\alpha = 0$.

4.12 The critical case

The validity of the method employed in the previous sections depends largely on the fact that we can map the x interval on the ξ interval (1–1). This fails if $\{r(x)/p(x)\}^{1/2}$ changes sign within the interval.

In the problems of interest $p(x)$ is bounded and vanishes, if at all, at the end points of the interval. This was the case in the example of the previous section in which $p(x)$ vanished only at $x = \pm 1$. We were then able to use the method in the interior of the interval, and in fact it was of little real embarrassment that the end points were excluded. We have detailed information about the behaviour of the solutions of (4.11.1) near $x = \pm 1$, which holds regardless of the magnitude of ν.

On the other hand if $r(x)$ changes sign we have a much more serious difficulty. In principle we could apply the previous methods to sub-intervals on either side of the zero, but the important question then arises of linking up the solutions on the two sides of the zero. This is a problem of analytic continuation.

It may be easier to appreciate the difficulty intuitively, if, instead of using the transformations of section 4.10, we use the one we gave in Chapter 3,

$$\xi = \int_0^x \frac{dt}{p(t)}, \quad (4.12.1)$$

which reduces the equation to

$$\frac{d^2y}{d\xi^2} + \{\lambda R(\xi) + Q(\xi)\}y = 0, \quad (4.12.2)$$

where $R(\xi) = p(x)r(x)$ and $Q(\xi) = p(x)q(x)$. Then near a zero of $r(x)$, though λ may be very large, $\lambda R(\xi)$ can be as small as we please.

Let us consider a very simple example. We must assume a number of results from the chapter on Bessel functions, so the reader may prefer to postpone reading this till he has read Chapter 9. We take the equation

$$\frac{d^2y}{d\xi^2} + \mu^2\xi y = 0, \tag{4.12.3}$$

and suppose for the moment that $\xi > 0$. The transformation $y(\xi) = \xi^{1/2}Y(\eta)$, where $\eta = \frac{2}{3}\xi^{3/2}$ (see exercise 9(a), Chapter 9) changes (4.12.3) to the form

$$\eta^2\frac{d^2Y}{d\eta^2} + \eta\frac{dY}{d\eta} + (\mu^2\eta^2 - \tfrac{1}{9})Y = 0. \tag{4.12.4}$$

The solutions of this can be written as linear combinations of $\mathcal{J}_{1/3}(\mu\eta)$ and $\mathcal{J}_{-1/3}(\mu\eta)$.

Since $\xi = 0$ is the critical point, we wish to avoid imposing boundary conditions there. However, we can see that we have two independent solutions of (4.12.3),

$$y_1(\xi) = \xi^{1/2}\mathcal{J}_{1/3}(\mu\eta) = \left(\frac{\mu}{3}\right)^{1/3}\xi\sum_{n=0}^{\infty}\frac{(-1)^n(\mu/3)^{2n}\xi^{3n}}{\Gamma(n+4/3)n!} \tag{4.12.5}$$

and

$$y_2(\xi) = \xi^{1/2}\mathcal{J}_{-1/3}(\mu\eta) = \left(\frac{\mu}{3}\right)^{-1/3}\sum_{n=0}^{\infty}\frac{(-1)^n(\mu/3)^{2n}\xi^{3n}}{\Gamma(n+2/3)n!}. \tag{4.12.6}$$

If $\xi > \delta > 0$ and μ is very large, we can apply the results for the asymptotic expansions of Bessel functions and deduce that

$$y_1(\xi) \sim \left(\frac{3}{\pi\mu}\right)^{1/2}\xi^{-1/4}\cos\left\{\tfrac{2}{3}\mu\xi^{3/2} - \frac{5\pi}{12}\right\}\left\{1 + O\left(\frac{1}{\mu\eta}\right)\right\} \tag{4.12.7}$$

and

$$y_2(\xi) \sim \left(\frac{3}{\pi\mu}\right)^{1/2}\xi^{-1/4}\cos\left\{\tfrac{2}{3}\mu\xi^{3/2} - \frac{\pi}{12}\right\}\left\{1 + O\left(\frac{1}{\mu\eta}\right)\right\}, \tag{4.12.8}$$

and these are clearly solutions that we could deduce from the analysis of section 4.10.

On the other hand when ξ is negative we can still apply (4.12.5) or (4.12.6) because the right-hand sides are entire functions. Thus

$$y_1(-|\xi|) = -\left(\frac{\mu}{3}\right)^{1/3}|\xi|\sum_{n=0}^{\infty}\frac{(\mu/3)^{2n}|\xi|^{3n}}{\Gamma(n+4/3)n!} = -|\xi|^{1/2}I_{1/3}(\mu\zeta) \tag{4.12.9}$$

and similarly

$$y_2(-|\xi|) = |\xi|^{1/2}I_{-1/3}(\mu\zeta), \tag{4.12.10}$$

where $\zeta = \frac{2}{3} | \xi |^{3/2}$.

If $\xi < -\delta < 0$ and μ is large,

$$y_1(- | \xi |) \sim - \left(\frac{3}{4\pi\mu}\right)^{1/2} | \xi |^{-1/4} \exp \{\tfrac{2}{3}\mu | \xi |^{2/3}\}\left\{1 + O\left(\frac{1}{| \mu\zeta |^2}\right)\right\} \tag{4.12.11}$$

with a similar result for $y_2(- | \xi |)$. In this case we can therefore actually perform the continuation.

Finally, consider equation (4.10.1) again. Suppose that $r(0) = 0$, that $r'(x) > 0$ in some interval including the origin and that in the same interval $p(x) > 0$. Then the transformations

$$y(x) = \{p(x)r(x)f(x)\}^{-1/4} Y(\xi), \qquad \xi = f(x),$$

where

$$2\{f(x)\}^{3/2} = 3\int_0^x \left\{\frac{r(t)}{p(t)}\right\}^{1/2} dt,$$

produce the equation

$$\frac{d^2Y}{d\xi^2} + \{\mu^2\xi + Q(\xi)\}Y = 0 \tag{4.12.12}$$

and we can map the interval round the origin (1–1) onto an interval which contains $\xi = 0$, so that origins correspond.

Then we can apply the results from our simple example to obtain the asymptotic behaviour of the solution, as we did before. For further details see the papers by LANGER [10–13].

EXERCISES

1. Find by successive integration by parts an asymptotic expansion for

$$\operatorname{erfc}(z) = \frac{2}{\sqrt{\pi}}\int_z^\infty \exp(-t^2)\, dt.$$

Define the sector in which the expansion is valid. Find also an estimate for the optimum error when $z = n$, a positive integer greater than unity.

2. Show that if $| z | > 1$, $a > 0$

$$\int_z^\infty \frac{e^{-at}\, dt}{(1 + t^2)^{1/2}} = E_1(az) + \sum_{s=1}^\infty (-1)^s A_s z^{-2s} E_{2s+1}(az),$$

where $A_s = (2s)!/2^{2s}(s!)^2$ and $E_\nu(z)$ is defined in section 4.2.

Hence or otherwise derive an asymptotic expansion for the integral on the left-hand side in the sector $| \arg z | \leqslant \pi/2 - \delta$ for $0 < \delta < \pi/2$.

3. Use Watson's lemma to find an asymptotic expansion for

$$\int_0^\infty \frac{e^{-zt}\, dt}{(1 + t^2)^{1/2}}$$

valid in the sector $| \arg z | \leqslant \pi/2 - \delta$, for $0 < \delta < \pi/2$.

4. Use Watson's lemma to find an asymptotic expansion for

$$\int_1^\infty \frac{e^{-zt}\,dt}{(t^2-1)^{1/2}}$$

valid in the sector $|\arg z| \leqslant \pi/2 - \delta$, for $0 < \delta < \pi/2$.

5. If $f(x)$ is regular in the neighbourhood of $x = 0$, and for small x

$$f(x) = \sum_{s=0}^{\infty} a_s x^s,$$

where $a_0 \neq 0$ and $\mu > 1$, show that if λ is large and positive,

$$\int_0^1 f(x)x^{\mu-1}(1-x)^{\lambda-1}\,dx \sim \frac{a_0\,\Gamma(\mu)}{\lambda^\mu} + \left(a_1 - \frac{\mu-1}{2}a_0\right)\frac{\Gamma(\mu+1)}{\lambda^{\mu+1}} + \ldots.$$

[*Hint*: put $y = -\log(1-x)$ and use Watson's lemma.]

6. Show that the point at infinity is an irregular singular point of the Laguerre equation

$$zw'' + (\alpha - z)w' + \nu w = 0$$

and obtain the first three terms of the asymptotic expansions of each of two independent solutions.

7. Show that the point at infinity is an irregular singular point of the equation

$$zw'' - 2aw' + b^2 zw = 0$$

(a and b real and positive) and obtain the first three terms of the asymptotic expansion of each of two independent solutions.

8. Show that the point at infinity is an irregular point of the equation

$$zw'' + (1 - 2az)w' + a^2 zw = 0, \qquad (a > 0)$$

and obtain the first three terms of the asymptotic expansion for each of two independent solutions.

9. Use the W.K.B. method to find the first three terms of the asymptotic expansions of a general solution of

$$\frac{d^2y}{dx^2} + (\mu^2 + x^2)y = 0$$

when $\mu \gg 1$.

10. Show that in an interval $[a, b]$, $r(x) > 0$, the W.K.B. asymptotic solutions of

$$\frac{d^2y}{dx^2} + \{\mu^2 r(x) + q(x)\}y = 0,$$

when $\mu \gg 1$, can be formally obtained by substituting into the equation expansions of the form

$$y(x) \sim \exp\left[\pm i\mu\int_0^x \{r(t)\}^{1/2}\,dt\right]\left[\sum_{n=0}^{\infty}\mu^{-n}f_n(x)\right],$$

and equating to zero the coefficients of like powers of μ.

[A similar technique could be applied to

$$\frac{d^2y}{dx^2} + \{\mu^2 r(x) + \mu s(x) + q(x)\}y = 0,$$

provided $r(x) > 0$, on $[a, b]$.]

ANSWERS

2. $\dfrac{e^{-az}}{z} \sum_{k=0}^{\infty} (-1)^k k! b_k (az)^{-k}$

with $b_{2m} = b_{2m+1} = 1 + \sum_{s=1}^{m} (-1)^s \dfrac{A_s}{(2s)!} a^{2s}$.

3. $z^{-1}\left\{1 + \sum_{s=1}^{\infty} (-1)^s a_s^2 z^{-2s}\right\}$ where $a_s = (2s)!/2^s(s!)$.

4. $\left(\dfrac{\pi}{2z}\right)^{1/2}\left\{1 + \sum_{s=1}^{\infty} (-1)^s (s!) A_s^2 (2z)^{-s}\right\}$, where $A_s = (2s)!/2^{2s}(s!)^2$.

6. $$w_1(z) \sim z^{\nu}\left\{1 - \frac{(-\nu)(1-\nu-\alpha)}{z} + \frac{(1-\nu)(-\nu)(1-\nu-\alpha)(2-\nu-\alpha)}{z^2} - \dots\right\}$$

and $$w_2(z) \sim z^{-(\alpha+\nu)} e^z \left\{1 + \frac{(\alpha+\nu)(1+\nu)}{z} + \frac{(1+\alpha+\nu)(\alpha+\nu)(1+\nu)(2+\nu)}{z^2} + \dots\right\}.$$

7. $$w(z) = e^{\pm ibz} z^a \left\{1 + \frac{(-a)(a+1)}{1!}\left(\pm\frac{1}{2ibz}\right) + \frac{(1-a)(-a)(a+1)(a+2)}{2!}\left(\pm\frac{1}{2ibz}\right)^2 + \dots\right\}.$$

8. $$w(z) = z^{-1/4} \exp [\{(az)^{1/2} \pm i\}^2]\left[1 \mp \frac{1}{1!\,16} i (az)^{-1/2} - \frac{1.9}{2!\,(16)^2}(az)^{-1} + \dots\right].$$

9. $$y(x) = \cos \alpha \cos \mu x - \mu^{-1} \{\sin \alpha + \tfrac{1}{6}x^3 \cos \alpha\} \sin \mu x - x^2 \cos \mu x\{(18 + x^4) \cos \alpha + 12x \sin \alpha\}/72\mu^2 + O(\mu^{-3}).$$

CHAPTER 5

The Gamma Function

Unlike the other functions to be considered in this volume, the subject of this chapter cannot be obtained as the solution of a differential equation. It is wanted primarily as an auxiliary function.

5.1 The integral definition

Let $z = x + iy$ be a complex number and consider the integral

$$\int_{\delta}^{T} e^{-t} t^{z-1} \, dt. \tag{5.1.1}$$

We define the branch of the many-valued function t^{z-1} by

$$t^{z-1} = \exp\{(z-1)\log t\}$$

where $\log t$ is real when $0 < t < \infty$. Thus

$$|t^{z-1}| = |\exp\{(z-1)\log t\}| = \exp\{(x-1)\log t\} = t^{x-1}.$$

If $T'' > T'$,

$$\left| \int_{T'}^{T''} e^{-t} t^{z-1} \, dt \right| \leqslant \int_{T'}^{T''} e^{-t} t^{x-1} \, dt.$$

Now e^{+t} increases faster than any power of t and so $e^{-t} < t^{-(x+1)}$, if $t > t_0$ for some t_0. If $T'' > T' > t_0$,

$$\int_{T'}^{T''} e^{-t} t^{x-1} \, dt < \int_{T'}^{T''} \frac{dt}{t^2} < \frac{1}{T'},$$

and the integral converges as $T' \to \infty$ for all complex z.

On the other hand if $0 < \delta' < \delta'' < \delta < 1$,

$$\int_{\delta'}^{\delta''} e^{-t} t^{x-1} \, dt = \int_{\delta'}^{\delta''} t^{x-1} \, dt\{1 + O(\delta)\} = \frac{(\delta'')^x - (\delta')^x}{x}\{1 + O(\delta)\}$$

and this remains bounded as $\delta' \to 0$ only if $x > 0$.

Thus the integral $\int_0^{\infty} e^{-t} t^{z-1} \, dt$, called the Euler integral of the second kind, which we shall denote by $\Gamma(z)$, converges absolutely for $\mathscr{R}(z) > 0$, and it converges uniformly for $X \geqslant \mathscr{R}(z) \geqslant \varepsilon > 0$, where X can be arbitrarily large.

Now consider the sequence of functions

$$f_n(z) = \int_0^n e^{-t} t^{z-1}\, dt. \tag{5.1.2}$$

Each integral converges absolutely and uniformly if $X \geqslant x \geqslant \varepsilon > 0$, and since the integrands are all regular functions of z, each function $f_n(z)$ is regular in the strip $\varepsilon \leqslant \mathscr{R}(z) \leqslant X$. This sequence, however, converges uniformly to $\Gamma(z)$, which therefore is regular in the strip. Since X is arbitrarily large, and ε is arbitrarily small, $\Gamma(z)$ represents a function which is regular in the whole right-hand half plane.

An immediate consequence of the definition is that if $x > 1$

$$\Gamma(z + 1) = \int_0^\infty e^{-t} t^z\, dt = -e^{-t} t^z \Big|_0^\infty + z \int_0^\infty e^{-t} t^{z-1}\, dt = z\,\Gamma(z). \tag{5.1.3}$$

In particular, if n is a positive integer,

$$\Gamma(n + 1) = n\,\Gamma(n) = n(n - 1)\,\Gamma(n - 1) = n!\,\Gamma(1) \tag{5.1.4}$$

and since

$$\Gamma(1) = \int_0^\infty e^{-t}\, dt = 1, \tag{5.1.5}$$

we have

$$\Gamma(n + 1) = n!. \tag{5.1.6}$$

This has led some authors to recommend that the notation $\Gamma(z + 1)$ should be replaced by $z!$.

By means of the transformation $t = u^\nu$ it follows from the definition integral that if $\nu > 0$

$$\Gamma\left(\frac{1}{\nu}\right) = \int_0^\infty e^{-t} t^{(1-\nu)/\nu}\, dt = \nu \int_0^\infty \exp(-u^\nu)\, du. \tag{5.1.7}$$

In particular,

$$\Gamma(\tfrac{1}{2}) = 2 \int_0^\infty \exp(-u^2)\, du. \tag{5.1.8}$$

5.2 The Euler products for $\Gamma(z)$

We require a simple inequality. If $0 < t < n$, then

$$0 \leqslant e^{-t} - (1 - t/n)^n \leqslant (t^2 e^{-t})/n. \tag{5.2.1}$$

To prove this, suppose that $0 < \xi < 1$; then

$$1 + \xi < e^\xi < 1 + \xi + \xi^2 + \xi^3 + \ldots = 1/(1 - \xi),$$

and so if $\xi = t/n$,

$$(1 + t/n)^n < e^t < (1 - t/n)^{-n}$$

or

$$(1 - t/n)^n < e^{-t} < (1 + t/n)^{-n}.$$

Thus

$$\begin{aligned} 0 < e^{-t} - (1 - t/n)^n &= e^{-t}\{1 - e^t(1 - t/n)^n\} \\ &\leqslant e^{-t}\{1 - (1 - t^2/n^2)^n\} \leqslant t^2\, e^{-t}/n. \end{aligned}$$

Now let

$$\gamma_n(z) = \int_0^n (1 - t/n)^n t^{z-1}\, dt.$$

Then

$$\begin{aligned} \Gamma(z) - \gamma_n(z) \,| &= \left| \int_0^n t^{z-1}\{e^{-t} - (1 - t/n)^n\}\, dt + \int_n^\infty e^{-t} t^{z-1}\, dt \right|, \\ &\leqslant \frac{1}{n}\int_0^n t^{x+1}\, e^{-t}\, dt + \int_n^\infty e^{-t} t^{x-1}\, dt, \end{aligned}$$

and since the first integral is certainly less than or equal to $\Gamma(x + 2)$ both terms tend to 0 as $n \to \infty$.

On the other hand, if we put $t = n\tau$,

$$\gamma_n(z) = n^z \int_0^1 (1 - \tau)^n \tau^{z-1}\, d\tau = n^z g_n(z), \tag{5.2.2}$$

say. If $x > 0$, we have

$$g_n(z) = \frac{1}{z}(1 - \tau)^n \tau^z \Big|_0^1 + \frac{n}{z}\int_0^1 (1 - \tau)^{n-1}\tau^z\, d\tau = \frac{n}{z} g_{n-1}(z + 1).$$

We may apply the recurrence relation n times to obtain

$$\begin{aligned} g_n(z) &= \frac{n!}{z(z + 1) \ldots (z + n - 1)} g_0(z + n) \\ &= \frac{n!}{z(z + 1) \ldots (z + n - 1)} \int_0^1 \tau^{n+z-1}\, d\tau \\ &= \frac{n!}{z(z + 1) \ldots (z + n - 1)(z + n)}. \end{aligned} \tag{5.2.3}$$

Finally we have

$$\Gamma(z) = \lim_{n\to\infty} \frac{n!\, n^z}{z(z + 1) \ldots (z + n - 1)(z + n)}. \tag{5.2.4}$$

An alternative product relation, also due to Euler, can be obtained in the form

$$\{\Gamma(z)\}^{-1} = \lim_{n\to\infty} zn^{-z}\prod_{k=1}^{n}(1 + z/k)$$
$$= \lim_{n\to\infty} z\prod_{k=1}^{n}\left(\frac{k+1}{k}\right)^{-z}\prod_{k=1}^{n}\left(1+\frac{z}{k}\right)$$
$$= z\prod_{k=1}^{\infty}\left\{\left(1+\frac{1}{k}\right)^{-z}\left(1+\frac{z}{k}\right)\right\}. \qquad (5.2.5)$$

5.3 The Weierstrass infinite product

We denote Euler's constant by γ, where

$$\gamma = \lim_{n\to\infty}\left\{\sum_{k=1}^{n}\frac{1}{k} - \log n\right\} = 0.5772157\ldots. \qquad (5.3.1)$$

Then

$$zn^{-z}\prod_{k=1}^{n}\left(1+\frac{z}{k}\right) = z\exp\left\{\left(\sum_{k=1}^{n}\frac{1}{k} - \log n\right)z\right\}\prod_{k=1}^{n}\left\{\left(1+\frac{z}{k}\right)e^{-z/k}\right\}.$$

If we now take the limit of both sides as $n \to \infty$, we derive the Weierstrass infinite product

$$\{\Gamma(z)\}^{-1} = ze^{\gamma z}\prod_{k=1}^{\infty}\{(1 + z/k)e^{-z/k}\}. \qquad (5.3.2)$$

We may choose an integer $k_0 > |z|$ corresponding to any arbitrary value of z. It follows that if $k > k_0$ we may expand $\log(1 + z/k)$ in powers of z/k. Hence

$$\log\left\{\left(1+\frac{z}{k}\right)e^{-z/k}\right\} = -\frac{z^2}{2k^2} + \frac{z^3}{3k^3} - \ldots$$

and so

$$\left|\log\left\{\left(1+\frac{z}{k}\right)e^{-z/k}\right\}\right| \leqslant \frac{|z|^2}{2k^2}\left\{1 + \frac{|z|}{k_0} + \frac{|z|^2}{k_0^{\,2}} + \ldots\right\}$$
$$= \frac{|z|^2}{2k^2}\left\{1 - \frac{|z|}{k_0}\right\}^{-1}.$$

Thus

$$\left|\sum_{k=k_0+1}^{\infty}\log\left\{\left(1+\frac{z}{k}\right)e^{-z/k}\right\}\right| \leqslant \frac{|z|^2}{2}\left\{1 - \frac{|z|}{k_0}\right\}^{-1}\left\{\sum_{k=k_0+1}^{\infty}\frac{1}{k^2}\right\}.$$

Hence the product (5.3.2) converges for every z, and so it represents an entire function of z with simple zeros at 0, -1, -2. . . . It therefore provides a continuation to the whole complex plane of $\{\Gamma(z)\}^{-1}$, which so far has only been defined for $\mathscr{R}(z) > 0$. It follows that $\Gamma(z)$ is a meromorphic function with simple poles at 0, -1, -2, . . .

The Weierstrass product gives convenient formulae for obtaining the derivatives of the gamma function. Since we have

$$\log \Gamma(z) = -\gamma z - \log z - \sum_{k=1}^{\infty} \left\{ \log\left(1 + \frac{z}{k}\right) - \frac{z}{k} \right\}, \tag{5.3.3}$$

then

$$\frac{\Gamma'(z)}{\Gamma(z)} = \frac{d}{dz} \log \Gamma(z) = -\gamma - \frac{1}{z} - \sum_{k=1}^{\infty} \left\{ \frac{1}{z+k} - \frac{1}{k} \right\} \tag{5.3.4}$$

and

$$\frac{d^2}{dz^2} \log \Gamma(z) = \sum_{k=0}^{\infty} \frac{1}{(z+k)^2}. \tag{5.3.5}$$

Higher derivatives may be similarly obtained.

5.4 The relation to sin πz

It follows from (5.3.2) that

$$\{\Gamma(z)\}^{-1}\{\Gamma(1-z)\}^{-1} = z(1-z)\, e^{\gamma} \prod_{k=1}^{\infty} \left\{ \left(1 + \frac{z}{k}\right) \left(1 + \frac{1-z}{k}\right) e^{-1/k} \right\}. \tag{5.4.1}$$

But

$$1 + \frac{1-z}{k} = \frac{k+1-z}{k} = \frac{k+1}{k}\left(1 - \frac{z}{k+1}\right)$$

and so

$$\prod_{k=1}^{n} \left(1 + \frac{1-z}{k}\right) = \frac{n+1}{1-z} \prod_{k=1}^{n+1} \left(1 - \frac{z}{k}\right). \tag{5.4.2}$$

Thus

$$\prod_{k=1}^{n} \left\{ \left(1 + \frac{z}{k}\right)\left(1 + \frac{1-z}{k}\right) e^{-1/k} \right\}$$
$$= \left(1 + \frac{z}{n+1}\right)^{-1} (1-z)^{-1}$$
$$\times \exp\left\{ \log(n+1) - \sum_{k=1}^{n} \frac{1}{k} \right\} \prod_{k=1}^{n+1} \left(1 - \frac{z^2}{k^2}\right).$$

In the limit as $n \to \infty$,

$$\{\Gamma(z)\}^{-1}\{\Gamma(1-z)\}^{-1} = z\prod_{k=1}^{\infty}\left(1-\frac{z^2}{k^2}\right) = S(z), \tag{5.4.3}$$

say. The analysis of the previous section will show that $S(z)$ is also an entire function of z. If we differentiate it logarithmically,

$$\frac{S'(z)}{S(z)} = \frac{1}{z} + \sum_{k=1}^{\infty}\frac{2z}{z^2-k^2} = \sum_{k=-\infty}^{\infty}\frac{1}{z-k}. \tag{5.4.4}$$

This series converges uniformly on any closed set excluding the poles $0, \pm 1, \pm 2, \ldots$ It clearly represents a function with the real period unity. At each simple pole k the residue is 1. Finally, for any fixed x, the right-hand side is bounded as $y \to \infty$. It has all of these properties in common with $\pi \cot \pi z$, and so, by the Liouville theorem, it can differ from $\pi \cot \pi z$ by an additive constant at most. Furthermore, since they both have the same principal parts near the origin, the constant is zero, and so

$$\frac{S'(z)}{S(z)} = \pi \cot \pi z \tag{5.4.5}$$

and it follows that $S(z) \propto \sin \pi z$. Near $z = 0$, $S(z) \sim z$, as is clear from the product, and so $S(z) = (\sin \pi z)/\pi$. Hence

$$\Gamma(z)\,\Gamma(1-z) = \pi \operatorname{cosec} \pi z. \tag{5.4.6}$$

This, combined with the definition of $\Gamma(z)$ in the right-hand half plane, gives the formal continuation of $\Gamma(z)$ to the rest of the complex plane. It is clear then that the recurrence relation

$$\Gamma(z+1) = z\,\Gamma(z) \tag{5.4.7}$$

may also be extended to the rest of the plane.

The residue of $\Gamma(z)$ at its simple pole at $-k$ is, on application of l'Hôpital's rule,

$$\lim_{z\to -k}(z+k)\,\Gamma(z) = \lim_{z\to -k}\frac{\pi(z+k)}{\sin \pi z}\,\frac{1}{\Gamma(1-z)} = \frac{(-1)^k}{k!}. \tag{5.4.8}$$

We see also from (5.4.6) that

$$\{\Gamma(\tfrac{1}{2})\}^2 = \pi \operatorname{cosec} \pi/2 = \pi, \tag{5.4.9}$$

and so from the integral relation of the first section

$$\int_0^{\infty} \exp(-u^2)\,du = \tfrac{1}{2}\sqrt{\pi}. \tag{5.4.10}$$

5.5 The beta function

The beta function $B(z, \zeta)$ is defined for $\mathscr{R}(z) > 0$, $\mathscr{R}(\zeta) > 0$ ($z = x + iy$, $\zeta = \xi + i\eta$) by

$$B(z, \zeta) = \int_0^1 t^{z-1}(1 - t)^{\zeta-1}\, dt, \tag{5.5.1}$$

which is known as the Eulerian integral of the first kind.

The proof that the integral so defined converges at both limits is essentially similar to the proof that the integral defining $\Gamma(z)$ converges near $t = 0$, if $\mathscr{R}(z) > 0$. We leave this as an exercise to the reader.

Since the substitution of $(1 - t)$ for t does not change the value of the integral, it is clearly a symmetric function of z and ζ. By putting $t = \cos^2 \theta$ we obtain the alternative form

$$B(z, \zeta) = 2\int_0^{\pi/2} \cos^{2z-1}\theta \sin^{2\zeta-1}\theta\, d\theta. \tag{5.5.2}$$

Let X be an arbitrarily large positive number. The substitutions $t = \tau^2$, $u = \sigma^2$ give

$$\left(\int_0^{X^2} e^{-t}t^{z-1}\, dt\right)\left(\int_0^{X^2} e^{-u}u^{\zeta-1}\, du\right) = 4\left\{\int_0^X \exp(-\tau^2)\tau^{2z-1}\, d\tau\right\} \times \left\{\int_0^X \exp(-\sigma^2)\sigma^{2\zeta-1}\, d\sigma\right\}.$$

These integrals converge absolutely and so can be replaced by the double integral

$$4\iint_{S_X} f(\tau, \sigma)\, d\tau\, d\sigma,$$

where

$$f(\tau, \sigma) = \tau^{2z-1}\sigma^{2\zeta-1} \exp\{-(\tau^2 + \sigma^2)\}$$

and S_X is the square domain $0 \leqslant \tau \leqslant X$, $0 \leqslant \sigma \leqslant X$. Let Q_X be the quadrant domain $0 \leqslant \tau^2 + \sigma^2 \leqslant X^2$, $\tau > 0$, $\sigma > 0$ and $S_{X/2}$ the square

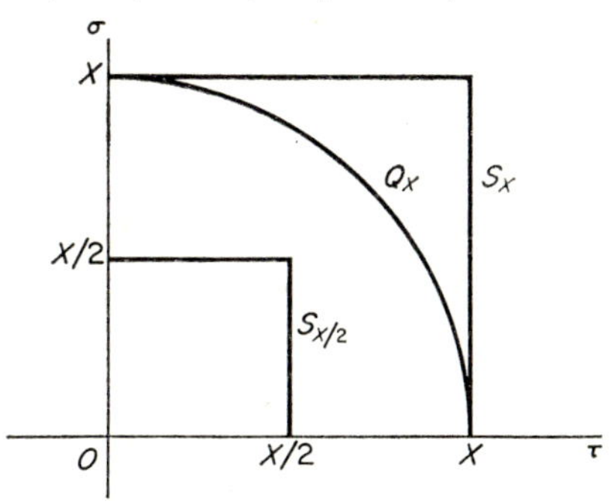

Fig. 5.1.

domain $0 \leqslant \tau \leqslant X/2$, $0 \leqslant \sigma \leqslant X/2$ (see Fig. 5.1). Then

$$\left| \iint_{S_X} f(\tau, \sigma)\, d\tau\, d\sigma - \iint_{Q_X} f(\tau, \sigma)\, d\tau\, d\sigma \right| = \left| \iint_{S_X - Q_X} f(\tau, \sigma)\, d\tau\, d\sigma \right|$$

$$\leqslant \iint_{S_X - Q_X} |f(\tau, \sigma)|\, d\tau\, d\sigma$$

$$= \iint_{S_X} |f(\tau, \sigma)|\, d\tau\, d\sigma - \iint_{Q_X} |f(\tau, \sigma)|\, d\tau\, d\sigma$$

$$\leqslant \iint_{S_X} |f(\tau, \sigma)|\, d\tau\, d\sigma - \iint_{S_{X/2}} |f(\tau, \sigma)|\, d\tau\, d\sigma.$$

Now as $X \to \infty$ the first integral on the right-hand side, from its original definition, is $\frac{1}{4}\Gamma(x)\,\Gamma(\zeta)$, and since also $X/2 \to \infty$, the second integral converges to the same sum. Now

$$\iint_{Q_X} f(\tau, \sigma)\, d\tau\, d\sigma = \left\{ \int_0^X \exp(-r^2) r^{2z+2\zeta-1}\, dr \right\} \times \left(\int_0^{\pi/2} \cos^{2z-1}\theta \sin^{2\zeta-1}\theta\, d\theta \right)$$

if we put $\tau = r\cos\theta$, $\sigma = r\sin\theta$. Finally, if $v = r^2$,

$$\Gamma(z)\,\Gamma(\zeta) = 2B(z,\zeta) \int_0^\infty \exp(-r^2) r^{2z+2\zeta-1}\, dr = B(z,\zeta) \int_0^\infty e^{-v} v^{z+\zeta-1}\, dv.$$

That is,

$$\Gamma(z)\,\Gamma(\zeta) = B(z, \zeta)\,\Gamma(z + \zeta). \tag{5.5.3}$$

5.6 The duplication formula

If we put $\zeta = z$ into (5.5.3) we obtain

$$\{\Gamma(z)\}^2 = \Gamma(2z) \int_0^1 t^{z-1}(1-t)^{z-1}\, dt = 2\,\Gamma(2z) \int_0^{1/2} t^{z-1}(1-t)^{z-1}\, dt. \tag{5.6.1}$$

The substitution

$$t = \tfrac{1}{2}(1 - \sqrt{\tau})$$

with

$$dt = -\tfrac{1}{4}\tau^{-1/2}\, d\tau \quad \text{and} \quad t(1-t) = \tfrac{1}{4} - (\tfrac{1}{2} - t)^2 = \tfrac{1}{4}(1 - \tau)$$

yields

$$\{\Gamma(z)\}^2 = 2^{1-2z}\,\Gamma(2z)\int_0^1 \tau^{1/2}(1-\tau)^{z-1}\,d\tau = 2^{1-2z}\,\Gamma(2z)B(\tfrac{1}{2}, z). \tag{5.6.2}$$

But from (5.5.3)

$$\Gamma(z+\tfrac{1}{2})B(\tfrac{1}{2}, z) = \Gamma(z)\,\Gamma(\tfrac{1}{2}) = \sqrt{\pi}\,\Gamma(z), \tag{5.6.3}$$

and so we arrive at the so-called duplication formula for the gamma function:

$$\sqrt{\pi}\,\Gamma(2z) = 2^{2z-1}\,\Gamma(z)\,\Gamma(z+\tfrac{1}{2}). \tag{5.6.4}$$

This can easily be continued to the rest of the complex plane, in analogy with the identities (5.1.3) and (5.4.6). It is worth remarking that if we put $2z = 2n + 1$, we get the special case

$$(2n)! = 2^{2n}(n!)\{(n-\tfrac{1}{2})(n-\tfrac{3}{2})\ldots(\tfrac{1}{2})\}. \tag{5.6.5}$$

5.7 The multiplication theorem

We now proceed to obtain a generalization of (5.6.4). It is easier to derive this directly from the derivative of the Weierstrass product (5.3.4), rather then from the beta function.

Let us put

$$\psi(z) = \Gamma'(z)/\Gamma(z) \tag{5.7.1}$$

and

$$-\ \psi_N(z) = \gamma + \frac{1}{z} + \sum_{k=1}^{N}\left(\frac{1}{z+k} - \frac{1}{k}\right). \tag{5.7.2}$$

Then

$$\begin{aligned} -\sum_{r=0}^{n-1}\psi_N\left(z+\frac{r}{n}\right) &= n\gamma + \sum_{r=0}^{n-1}\frac{n}{nz+r} + \sum_{k=1}^{N}\left(\sum_{r=0}^{n-1}\frac{n}{nz+r+nk} - \frac{n}{k}\right) \\ &= n\left\{\gamma + \frac{1}{nz} + \sum_{k=1}^{nN+n-1}\left(\frac{1}{nz+k} - \frac{1}{k}\right) + \sum_{k=N+1}^{nN+n-1}\frac{1}{k}\right\} \\ &= n\left\{-\psi_{nN+n-1}(nz) + \sum_{k=N+1}^{nN+n-1}\frac{1}{k}\right\}. \end{aligned} \tag{5.7.3}$$

If N is large enough,

$$\begin{aligned} \sum_{k=N+1}^{nN+n-1}\frac{1}{k} &= \gamma + \log(nN+n-1) - \gamma - \log(N+1) + O\left(\frac{1}{N}\right) \\ &= \log n + O\left(\frac{1}{N}\right). \end{aligned}$$

In the limit, therefore, as $N \to \infty$

$$\sum_{r=0}^{n-1} \psi(z + r/n) = n\psi(nz) - n \log n \tag{5.7.4}$$

and integrating this relation with respect to z, for some constant A_n, gives us

$$\prod_{r=0}^{n-1} \Gamma(z + r/n) = A_n n^{-nz}\, \Gamma(nz). \tag{5.7.5}$$

This holds for all z, and in particular we may put $z = 1/n$, so that

$$A_n = n \prod_{r=0}^{n-1} \Gamma\{(r + 1)/n\} \tag{5.7.6}$$

and is positive. In the product there are n terms of which the last is $\Gamma(1) = 1$, the remaining being $\Gamma(1/n)$, $\Gamma(2/n)$, . . . $\Gamma\{(n - 1)/n\}$. Hence from (5.4.6)

$$A_n{}^2 = n^2 \left\{ \Gamma\left(\frac{1}{n}\right) \Gamma\left(\frac{n-1}{n}\right) \right\} \left\{ \Gamma\left(\frac{2}{n}\right) \Gamma\left(\frac{n-2}{n}\right) \right\} \dots \times \left\{ \Gamma\left(\frac{n-1}{n}\right) \Gamma\left(\frac{1}{n}\right) \right\}$$

$$= n^2 \pi^{n-1} \operatorname{cosec} \frac{\pi}{n} \operatorname{cosec} \frac{2\pi}{n} \dots \operatorname{cosec} \frac{(n-1)\pi}{n}. \tag{5.7.7}$$

Now

$$\zeta^n - 1 = (\zeta - 1) \prod_{k=1}^{n-1} (\zeta - \alpha^k) \tag{5.7.8}$$

where $\alpha = \exp(2\pi \mathrm{i}/n)$. If we differentiate (5.7.8) we get

$$n\zeta^{n-1} = \prod_{k=1}^{n-1} (\zeta - \alpha^k) + (\zeta - 1) \frac{\mathrm{d}}{\mathrm{d}\zeta} \left\{ \prod_{k=1}^{n-1} (\zeta - \alpha^k) \right\}, \tag{5.7.9}$$

and so

$$n = \prod_{k=1}^{n-1} (1 - \alpha^k). \tag{5.7.10}$$

But

$$1 - \alpha^k = 1 - \exp(2\pi \mathrm{i}k/n) = \exp(\pi \mathrm{i}k/n)\{ \exp(-\pi \mathrm{i}k/n) - \exp(\pi \mathrm{i}k/n)\}$$

$$= -2\mathrm{i} \exp(\pi \mathrm{i}k/n) \sin(\pi k/n). \tag{5.7.11}$$

Since also

$$\prod_{k=1}^{n-1} \exp(\pi \mathrm{i}k/n) = \exp\{\pi \mathrm{i}n(n - 1)/2n\} = \{ \exp(\pi \mathrm{i}/2)\}^{n-1}$$

it follows from (5.7.10) and (5.7.11) that

$$n = 2^{n-1} \prod_{k=1}^{n-1} \sin(\pi k/n). \tag{5.7.12}$$

Finally, therefore we obtain

$$A_n = (2\pi)^{(n-1)/2} n^{1/2}. \tag{5.7.13}$$

5.8 The Hankel contour

It has been shown above that though the intitial integral $\int_0^\infty e^{-t} t^{z-1}\, dt$ defines $\Gamma(z)$ only in the half-plane $x > 0$, the function can be continued into the rest of the complex plane from which its simple poles are omitted. This can be accomplished by the relation $\Gamma(z)\,\Gamma(1 - z) = \pi \operatorname{cosec} \pi z$. It is nevertheless convenient to have an integral definition which is convergent in the complex plane.

Consider

$$F(z) = \int_C e^{\zeta} \zeta^{z-1}\, d\zeta, \tag{5.8.1}$$

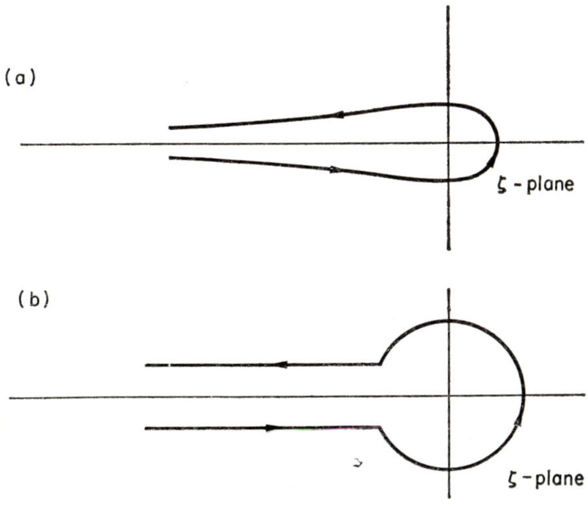

Fig. 5.2.

where C is the contour shown in Fig. 5.2(a), starting from ∞ on the negative real axis, going round the origin in the positive direction, and returning to its starting point. The integrand is many-valued, but we can make it single-valued by cutting the ζ-plane along the negative real axis and taking the principal branch of ζ^{z-1}.

For every ζ in the cut plane the integrand is an entire function of z, and the integral converges uniformly for every z in the closed disc

$|z| \leqslant R$ for an arbitrarily large R. Hence $F(z)$ is also an entire function of z.

Suppose now that $x > 0$. The contour can be distorted into that shown in Fig. 5.2(b), the two sides of the cut along the negative real axis for $\delta \leqslant |\zeta| < \infty$ and the broken circle $|\zeta| = \delta$, $-\pi < \arg \zeta \leqslant \pi$. If, then, we put

$$\zeta = t\mathrm{e}^{\mathrm{i}\phi},$$

$$\begin{aligned} F(z) &= \int_{\infty}^{\delta} \exp\{-t + (z-1)\log t - \mathrm{i}z\pi\}\,\mathrm{d}t \\ &\quad + \mathrm{i}\delta \int_{-\pi}^{\pi} \exp\{\delta\mathrm{e}^{\mathrm{i}\phi} + (z-1)\log\delta + \mathrm{i}z\phi\}\,\mathrm{d}\phi \\ &\quad + \int_{\delta}^{\infty} \exp\{-t + (z-1)\log t + \mathrm{i}z\pi\}\,\mathrm{d}t \\ &= I_1 + I_2 + I_3. \end{aligned} \tag{5.8.2}$$

Now

$$|I_2| \leqslant 2\pi\mathrm{e}^{\pi|y|}\delta^x\{1 + O(\delta)\} \tag{5.8.3}$$

and so I_2 vanishes with δ, since $x > 0$. Thus if we let $\delta \to 0$

$$F(z) = (\mathrm{e}^{\mathrm{i}z\pi} - \mathrm{e}^{-\mathrm{i}z\pi}) \int_0^{\infty} \mathrm{e}^{-t}t^{z-1}\,\mathrm{d}t = 2\mathrm{i}\sin\pi z\,\Gamma(z). \tag{5.8.4}$$

We may therefore define $\Gamma(z)$ in the whole complex plane as a quotient of entire functions by

$$\Gamma(z) = \frac{1}{2\mathrm{i}}\operatorname{cosec}\pi z \int_C \mathrm{e}^{\zeta}\zeta^{z-1}\,\mathrm{d}\zeta. \tag{5.8.5}$$

If we replace $z - 1$ by $-z$ and use (5.4.6), we see that

$$\{\Gamma(z)\}^{-1} = \frac{1}{2\pi\mathrm{i}} \int_C \mathrm{e}^{\zeta}\zeta^{-z}\,\mathrm{d}\zeta. \tag{5.8.6}$$

5.9 Stirling's approximation

Consider the sequence

$$u_n = n!\,\mathrm{e}^n n^{-n-1/2}. \tag{5.9.1}$$

Then

$$\begin{aligned} \frac{u_{n+1}}{u_n} &= \mathrm{e}\left(1 + \frac{1}{n}\right)^{-n-1/2} = \exp\left\{1 - (n + \tfrac{1}{2})\log\left(1 + \frac{1}{n}\right)\right\} \\ &= \exp\left\{-\frac{1}{12n^2}\right\}\left\{1 + O\left(\frac{1}{n^3}\right)\right\} < 1. \end{aligned} \tag{5.9.2}$$

Thus the sequence consists of positive terms and is strictly decreasing. It therefore tends to a limit, positive or zero. It also follows that

$$\log u_n - \log u_{n+1} < \frac{K}{n^2}$$

for some $K > 0$, and so

$$\log u_n - \log u_{n+m} < K \sum_{r=n}^{n+m} \frac{1}{r^2}.$$

However, $\sum_{r=1}^{\infty} (1/r^2)$ is finite, and so for a fixed n, $\log u_{n+m}$ is finite, and the limit of u_{n+m} cannot be zero. Therefore for some $C > 0$

$$\lim_{n\to\infty} u_n = C. \tag{5.9.3}$$

Now consider

$$\begin{aligned} U_k(z) &= \tfrac{1}{2} \int_0^1 t(t-1)(t+k+z)^{-2}\, dt \\ &= \tfrac{1}{2} \int_k^{k+1} \frac{(t-k)(t-k-1)}{(t+z)^2}\, dt \end{aligned} \tag{5.9.4}$$

where k is a non-negative integer, and the z-plane is cut along the negative real axis from $-\infty$ to 0.

Since

$$\begin{aligned} t(t-1) = (t+k+z)^2 - \{2(k+z)+1\}(t+k+z) \\ + (k+z)(k+z+1), \end{aligned} \tag{5.9.5}$$

we obtain a closed expression for (5.9.4), namely

$$\begin{aligned} U_k(z) = 1 + (k+z)\log(k+z) - (k+z+1)\log(k+z+1) \\ + \tfrac{1}{2}\log(k+z) + \tfrac{1}{2}\log(k+z+1). \end{aligned} \tag{5.9.6}$$

It follows that

$$\begin{aligned} \sum_{k=0}^{n-1} U_k(z) = n + z\log z - (n+z)\log(n+z) + \log\{z(z+1)\ldots \\ \times (z+n)\} - \tfrac{1}{2}\log z - \tfrac{1}{2}\log(n+z). \end{aligned} \tag{5.9.7}$$

If we define

$$\Gamma_n(z) = \frac{n!\, n^z}{z(z+1)\ldots(z+n)}, \tag{5.9.8}$$

then

$$\begin{aligned} \sum_{k=0}^{n-1} U_k(z) = (z-\tfrac{1}{2})\log z - \log\Gamma_n(z) + \log u_n + z\log n \\ + (n+\tfrac{1}{2})\log n - (n+z+\tfrac{1}{2})\log(z+n). \end{aligned} \tag{5.9.9}$$

If we fix z in the cut plane and take $n > |z|$, we may express this in the form

$$\sum_{k=0}^{n-1} U_k(z) = (z - \tfrac{1}{2}) \log z - \log \Gamma_n(z) + \log u_n - z + O\left(\frac{|z|^2}{n}\right). \tag{5.9.10}$$

On the other hand,

$$\sum_{k=0}^{n-1} U_k(z) = \int_0^n \frac{\phi(t)\,dt}{(t+z)^2}. \tag{5.9.11}$$

In the interval $(k, k+1)$ the function $\phi(t)$ has the form

$$\tfrac{1}{2}(t-k)(t-k-1) = \tfrac{1}{2}(t-k-\tfrac{1}{2})^2 - \tfrac{1}{8},$$

and so

$$|\phi(t)| \leqslant \tfrac{1}{8}. \tag{5.9.12}$$

Thus if $|\arg z| \leqslant \pi - \delta$,

$$\begin{aligned}\left|\sum_{k=0}^{n} U_k(z)\right| &\leqslant \tfrac{1}{8}\int_0^\infty \frac{dt}{(t+x)^2 + y^2} \\ &= \frac{1}{8y}\tan^{-1}\left(\frac{y}{x}\right) \leqslant \frac{\pi}{16\,|z|\sin\delta}.\end{aligned} \tag{5.9.13}$$

If then we let $n \to \infty$,

$$\Gamma(z) = Cz^{z-1/2}e^{-z}\left\{1 + O\left(\frac{1}{|z|}\right)\right\} \tag{5.9.14}$$

in $|\arg z| \leqslant \pi - \delta$, for every $\delta > 0$.

To find the constant C we use the duplication formula in the form

$$\log \Gamma(2z) + \tfrac{1}{2}\log \pi = (2z-1)\log 2 + \log \Gamma(z) + \log \Gamma(z + \tfrac{1}{2}). \tag{5.9.15}$$

We replace the gamma functions by means of (5.9.14) and so derive

$$\begin{aligned}\log C + (2z - \tfrac{1}{2})\log 2 + (2z - \tfrac{1}{2})\log z - 2z + O\left(\frac{1}{|z|}\right) + \tfrac{1}{2}\log \pi \\ = (2z-1)\log 2 + \log C + (z - \tfrac{1}{2})\log z - z \\ + \log C + z\log(z + \tfrac{1}{2}) - z - \tfrac{1}{2} + O\left(\frac{1}{|z|}\right).\end{aligned} \tag{5.9.16}$$

We may assume that $|z| > 1$, so that

$$z \log(z + \tfrac{1}{2}) = z\log z + \tfrac{1}{2} + O\left(\frac{1}{|z|}\right) \tag{5.9.17}$$

and it follows that

$$C = \sqrt{(2\pi)}. \tag{5.9.18}$$

We have established, therefore, Stirling's formula;

$$\log \Gamma(z) = \tfrac{1}{2}\log(2\pi) + (z - \tfrac{1}{2})\log z - z + h(z) \tag{5.9.19}$$

or

$$\Gamma(z) = \left(\frac{2\pi}{z}\right)^{1/2} z^z \exp\{-z + h(z)\}, \tag{5.9.20}$$

valid for $|z| > 1$ and $|\arg z| \leqslant \pi - \delta > 0$.

We can estimate the error term if we integrate

$$\int_0^n \frac{\phi(t)\,dt}{(t+z)^2}$$

twice by parts. This will replace the final term in (5.9.19) by

$$h(z) = \frac{1}{12z} + O\left(\frac{1}{|z|^3}\right). \tag{5.9.21}$$

This could be further improved by integrating again, but the labour involved soon becomes prohibitive. Thus if we wish to derive a complete asymptotic expansion for $\Gamma(z)$ we must use other methods.

5.10 Binet's first integral

From (5.3.4) we have

$$\begin{aligned}\frac{\Gamma'(z)}{\Gamma(z)} &= -\gamma - \frac{1}{z} - \sum_{k=1}^{\infty}\left\{\frac{1}{z+k} - \frac{1}{k}\right\}\\ &= \lim_{n\to\infty}\left\{\log n - \sum_{k=0}^{n}\frac{1}{z+k}\right\} = \lim_{n\to\infty}\phi_n(z), \text{ say.}\end{aligned} \tag{5.10.1}$$

If $\mathscr{R}(z+k) > 0$,

$$(z+k)^{-1} = \int_0^\infty \exp\{-(z+k)t\}\,dt,$$

and if $\mathscr{R}(\alpha) > 0$,

$$\begin{aligned}\int_0^\infty \frac{e^{-t} - e^{-t\alpha}}{t}\,dt &= \lim_{\varepsilon\to 0}\int_\varepsilon^\infty \frac{e^{-t} - e^{-\alpha t}}{t}\,dt\\ &= \lim_{\varepsilon\to 0}\left(\int_\varepsilon^\infty \frac{e^{-t}}{t}\,dt - \int_{\alpha\varepsilon}^\infty \frac{e^{-u}}{u}\,du\right)\\ &= \lim_{\varepsilon\to 0}\int_\varepsilon^{\alpha\varepsilon}\frac{e^{-t}}{t}\,dt = \log\alpha.\end{aligned}$$

Hence if $\mathscr{R}(z) > 0$,

$$\begin{aligned}\phi_n(z) &= \int_0^\infty\left(\frac{e^{-t} - e^{-nt}}{t} - \sum_{k=0}^{n} e^{-(z+k)t}\right)dt,\\ &= \int_0^\infty\left\{\frac{e^{-t} - e^{-nt}}{t} - e^{-zt}\left(\frac{1 - e^{-nt}}{1 - e^{-t}}\right)\right\}dt,\\ &= \int_0^\infty\left(\frac{e^{-t}}{t} - \frac{e^{-zt}}{1 - e^{-t}}\right)dt + R_n(z),\end{aligned}$$

where

$$R_n(z) = \int_0^\infty e^{-nt}\left(\frac{e^{-zt}}{1 - e^{-t}} - \frac{1}{t}\right) dt.$$

The term in brackets tends to $-z$ as $t \to 0$, is continuous for every $t > 0$, and tends uniformly to zero as $t \to \infty$. Hence for all t in the interval $0 \leqslant t < \infty$,

$$\left|\left\{\frac{e^{-zt}}{1 - e^{-t}} - \frac{1}{t}\right\}\right| \leqslant M(z)$$

where M depends on z but not on t. Then

$$|R_n(z)| \leqslant M(z) \int_0^\infty e^{-nt}\, dt = M(z)/n$$

and so tends to zero as $n \to \infty$.

We therefore deduce that

$$\frac{\Gamma'(z)}{\Gamma(z)} = \int_0^\infty \left(\frac{e^{-t}}{t} - \frac{e^{-zt}}{1 - e^{-t}}\right) dt. \tag{5.10.2}$$

We have shown in the previous section that

$$\log \Gamma(z) = \tfrac{1}{2}\log 2\pi + (z - \tfrac{1}{2})\log z - z + h(z) \tag{5.10.3}$$

and so

$$\frac{\Gamma'(z)}{\Gamma(z)} = \log z - \frac{1}{2z} + h'(z).$$

If we use results similar to those we have applied above when $\mathscr{R}(z) > 0$, the right-hand side can be written

$$\int_0^\infty \left\{\frac{e^{-t} - e^{-tz}}{t} - \tfrac{1}{2}e^{-tz}\right\} dt + h'(z). \tag{5.10.4}$$

We subtract (5.10.4) from (5.10.2) and derive

$$h'(z) = \int_0^\infty e^{-tz}\left\{\frac{1}{t} + \frac{1}{2} - \frac{1}{1 - e^{-t}}\right\} dt. \tag{5.10.5}$$

The term in brackets is $O(t)$ as $t \to 0$, is continuous for all positive t, and tends to $-\frac{1}{2}$ as $t \to \infty$. It is therefore bounded. We may therefore integrate (5.10.5) under the integral sign and so derive

$$h(z) = \text{const.} + \int_0^\infty \frac{e^{-tz}}{t}\left(\frac{1}{t} + \frac{1}{2} - \frac{1}{1 - e^{-t}}\right) dt.$$

The term multiplying e^{-tz} is still bounded near $t = 0$, and so is continuous on $0 \leqslant t < \infty$ and tends to zero as $t \to 0$. It is thus absolutely bounded, and so for some K

$$\left|\int_0^\infty \frac{e^{-tz}}{t}\left(\frac{1}{t} + \frac{1}{2} - \frac{1}{1 - e^{-t}}\right)\right| dt \leqslant K \int_0^\infty e^{-tx}\, dt = K/x$$

and so tends to zero as $x \to \infty$. We have shown in the previous section that $h(z) \to 0$ as $|z| \to \infty$ in $|\arg z| \leqslant \pi - \delta$. It follows that the constant of integration is zero. Then

$$h(z) = \int_0^\infty \frac{e^{-tz}}{t}\left(\frac{1}{t} + \frac{1}{2} - \frac{1}{1 - e^{-t}}\right) dt. \tag{5.10.6}$$

Since

$$\frac{1}{1 - e^{-t}} = \frac{e^t}{e^t - 1} = 1 + \frac{1}{e^t - 1},$$

we may write this in its more usual form, known as Binet's first integral:

$$h(z) = \int_0^\infty \frac{e^{-tz}}{t}\left(\frac{1}{t} - \frac{1}{2} - \frac{1}{e^t - 1}\right) dt. \tag{5.10.7}$$

5.11 The asymptotic series for log $\Gamma(z)$

We have remarked in the previous section that

$$\frac{1}{t} - \frac{1}{2} - \frac{1}{e^t - 1}$$

is regular near $t = 0$, vanishes like t there and is absolutely bounded for all real positive t. The term multiplying e^{-tz} in (5.10.7) therefore satisfies all the necessary conditions for the application of Watson's lemma.

The Bernoulli numbers are defined by

$$\frac{t}{e^t - 1} = 1 - \frac{t}{2} + \sum_{s=1}^{\infty} (-1)^{s-1} \frac{B_s t^{2s}}{(2s)!}.$$

If we then write

$$h(z) = \int_0^\infty e^{-tz} f(t)\, dt, \tag{5.11.1}$$

it follows that near $t = 0$

$$f(t) = \sum_{s=0}^{\infty} (-1)^{s-1} \frac{B_{s+1} t^{2s}}{(2s+2)!}. \tag{5.11.2}$$

Hence Watson's lemma gives

$$h(z) \sim \sum_{s=0}^{\infty} (-1)^{s-1} \frac{B_{s+1}\, z^{-2s-1}}{(2s+1)(2s+2)}. \tag{5.11.3}$$

within the sector $|\arg z| \leqslant \pi/2 - \delta$.

5.12 The continuation of the asymptotic series

We saw in section 5.9 that the error in the Stirling formula was uniformly of order $1/|z|$ in the larger sector $|\arg z| \leqslant \pi - \delta$. It therefore seems reasonable that the expansion (5.11.3) should also be valid in the larger sector.

The restriction $\mathcal{R}(z) > 0$ was necessary in the analysis of section 5.10 to ensure that the integrals with the term e^{-tz} converge. We have also applied the Watson lemma in the previous section. This had a similar restriction, but in section 4.4 we were only able to show that it gave an asymptotic series in the closed sector $|\arg z| \leqslant \pi/2 - \delta$.

If in section 5.10 we used the same basic integrals, but tilted the line of integration through an angle α, the new requirement would be that

$$\mathcal{R}(ze^{i\alpha}) > 0,$$

or what is the same thing,

$$-\frac{\pi}{2} - \alpha < \arg z < \frac{\pi}{2} - \alpha.$$

There is a limit to the amount of tilt as we also require that integrals involving e^{-t} converge and we must avoid the singularities of $(e^t - 1)^{-1}$, all of which lie on the imaginary axis. We are therefore restricted to

$$-\frac{\pi}{2} + \varepsilon \leqslant \alpha \leqslant \frac{\pi}{2} - \varepsilon.$$

We may establish the validity of Watson's lemma also with the line of integration tilted.

The conclusion we derive is that we may establish the validity of (5.11.3) in a half plane $-\pi/2 - \alpha < \arg z < \pi/2 - \alpha$, for every α in the range $-\pi/2 + \varepsilon \leqslant \alpha \leqslant \pi/2 - \varepsilon$. Since these half planes overlap each other, the result must be analytically valid in the union of the domains, which is the whole plane with a sector enclosing the negative real axis removed.

Exercises

1. If a, b, c and d are all positive and $a + b = c + d$ and if

$$K_n = \frac{\Gamma(n+a)\,\Gamma(n+b)}{\Gamma(n+c)\,\Gamma(n+d)},$$

show that as $n \to \infty$, $K_n \to 1$.

2. Use the first Euler product for $\Gamma(z)$, combined with the result of exercise 1, to establish the triplication formula

$$2\pi\,\Gamma(3z) = 3^{3z-1/2}\,\Gamma(z)\,\Gamma(z+\tfrac{1}{3})\,\Gamma(z+\tfrac{2}{3}), \qquad (\Gamma(\tfrac{1}{3})\,\Gamma(\tfrac{2}{3}) = 2\pi/\sqrt{3}).$$

3. Use the Weierstrass product for $\Gamma(z)$ to prove that

$$\Gamma'(1) = -\gamma,$$

where γ is Euler's constant. Write down an expression for $\Gamma'(n)$ where n is a positive integer.

4. If the Laplace transform of a function $f(t)$, when it exists, is defined by

$$\int_0^\infty e^{-st} f(t)\, dt,$$

find the Laplace transforms of t^ν and $t^\nu \log t$, where $\mathscr{R}(\nu) > -1$.

5. Let C_1 be the contour consisting of the line segments $\delta \leqslant x \leqslant R$, $\delta \leqslant y \leqslant R$, and the quadrants $|z| = \delta$, $0 \leqslant \arg z \leqslant \pi/2$, and $|z| = R$, $0 \leqslant \arg z \leqslant \pi/2$, and C_2 its mirror image in the positive real axis. By considering

$$\int_{C_1} e^{iz} z^\nu\, dz \quad \text{and} \quad \int_{C_2} e^{-iz} z^\nu\, dz,$$

where $-1 < \mathscr{R}(\nu) < 0$, show that

$$\int_0^\infty x^\nu \cos x\, dx = \Gamma(\nu + 1) \cos \tfrac{1}{2}(\nu + 1)\pi$$

and

$$\int_0^\infty x^\nu \sin x\, dx = \Gamma(\nu + 1) \sin \tfrac{1}{2}(\nu + 1)\pi.$$

Deduce that the second result, but not the first, holds if $-2 < \mathscr{R}(\nu) < 0$.

6. Integrate e^{-z^n} round the sector $\arg z = 0$, $0 \leqslant |z| \leqslant R$, $0 \leqslant \arg z \leqslant \alpha$, $|z| = R$, and $\arg z = \alpha$, $R \geqslant |z| \geqslant 0$, where $0 < \alpha \leqslant \pi/2n$, and n is an integer $\geqslant 2$. Hence deduce that

$$n \int_0^\infty \exp\{-r^n \cos n\alpha\} \cos\{r^n \sin n\alpha\} = \Gamma(1/n) \cos \alpha.$$

7. Show that if $0 < \mathscr{R}(z) < 1$

$$\Gamma(z)\,\Gamma(1 - z) = \iint e^{-(t+u)} \left(\frac{t}{u}\right)^z \frac{dt\, du}{t},$$

where the integral is to be taken over the whole of the first quadrant of the (t, u)-plane. By defining new coordinates $\xi = t + u$, $\eta = t/u$, prove that

$$\Gamma(z)\,\Gamma(1 - z) = \frac{\pi}{\sin \pi z}.$$

8. If x is real and positive, show that

$$|\Gamma(ix)| = \left(\frac{\pi}{x \sinh \pi x}\right)^{1/2}.$$

9. Show that $\Gamma(z)$ has no finite zeros.

10. Show that

(a) $$\int_0^1 x^a (1 - x^b)^{c-1}\, dx = \frac{1}{b} B\left(\frac{a+1}{b}, c\right), \qquad \mathscr{R}\left(\frac{a+1}{b}\right) > 0, \mathscr{R}(c) > 0.$$

(b) $$\int_0^\infty \frac{x^a}{(1 + x)^b}\, dx = B(a + 1, b - a - 1), \qquad a > -1, \quad b > a + 1.$$

[*Hint:* put $y = x/(1 + x)$.]

(c) Show that if n is an integer, and $\nu > -1$,

$$\int_{-1}^{1} x^n(1 - x^2)^\nu \, dx = 0 \qquad \text{if } n \text{ is odd,}$$

$$= B\left(\frac{n-1}{2}, \nu + 1\right) \quad \text{if } n \text{ is even.}$$

11. If C is the double-dumb-bell contour about the points $t = 0$ and $t = 1$ (see Fig. 2.6) and $\mathscr{R}(z) > 0, \mathscr{R}(\zeta) > 0$, show that

$$\int_C t^{z-1}(1 - t)^{\zeta-1} \, dt = -4e^{i\pi(z+\zeta)} \sin \pi z \sin \pi\zeta B(z, \zeta).$$

Hence write down a formal definition of $B(z, \zeta)$, valid for all finite complex values of z and ζ except for the negative integers.

12. If $a > 0$, $b > a$, and n is a positive integer, express $\Gamma(a + n)/\Gamma(b + n)$ in terms of $B(a + n, b - a)$. Hence use the result of exercise 5 of Chapter 4 to show that as $n \to \infty$,

$$\frac{\Gamma(a+n)}{\Gamma(b+n)} \sim n^{a-b}\left\{1 - \frac{(b-a)(b+a-1)}{2n} + O\left(\frac{1}{n^2}\right)\right\}.$$

Verify that this result could be obtained directly from Stirling's formula.

13. If $a > 0, b > 0, c > 0$ and $A_n = \{\Gamma(a + n)\,\Gamma(b + n)\}/\{\Gamma(c + n)\,\Gamma(1 + n)\}$, use the previous example to show that $A_n \sim n^{a+b-c-1}$. Deduce that if $c > a + b$, the series $\sum_{n=0}^{\infty} A_n z^n$ converges at every point of $|z| = 1$.

14. If $F(z) = \{2^{2z-1}\,\Gamma(z)\,\Gamma(z + \tfrac{1}{2})\}/\Gamma(2z)$, show that $F(z)$ is an entire function of z, and that for every z

$$F(z + 1) = F(z).$$

By considering $d^2 \log F(z)/dz^2$ or otherwise show that $F(z)$ is a constant, and hence deduce the duplication formula.

15. Let C_k^+ be the rectangle with vertices at k, $k + 1$, $k + 1 + iX$, $k + iX$, indented to exclude the points k and $k + 1$, and let C_k^- be the mirror image of C_k^+ in the positive real axis. Let also

$$I_k^+ = \int \frac{d\zeta}{(\zeta + z)^2(e^{-2\pi i\zeta} - 1)}$$

where the contour of integration is C_k^+ and

$$I_k^- = \int \frac{d\zeta}{(\zeta + z)^2(e^{2\pi i\zeta} - 1)}$$

where the contour of integration is C_k^-. Then show that if $\mathscr{R}(z) > 0$

$$\frac{d^2}{dz^2}\{\log \Gamma(z)\} = \sum_{k=0}^{\infty} \frac{1}{(z+k)^2} = \frac{1}{z} + \frac{1}{2z^2} + 4z\int_0^\infty \frac{\eta \, d\eta}{(e^{2\pi\eta} - 1)(z^2 + \eta^2)^2}.$$

Hence or otherwise deduce that if $\mathscr{R}(z) > 0$, then

$$h(z) = 2\int_0^\infty \frac{\tan^{-1}(\eta/z)\, d\eta}{e^{2\pi\eta} - 1},$$

(Binet's second integral).

16. Given the expression for the Bernoulli numbers

$$B_n = 4n \int_0^\infty \frac{\eta^{2n-1}\,\mathrm{d}\eta}{\mathrm{e}^{2\pi\eta} - 1},$$

deduce the asymptotic expansion of $\Gamma(z)$ in the half-plane $\mathscr{R}(z) > 0$ from Binet's second integral.

ANSWERS

3. $\Gamma'(n) = (n-1)!\left\{\sum_{k=1}^{n-1} \frac{1}{k} - \gamma\right\}.$

4. $\Gamma(\nu+1)s^{-(1+\nu)}, \quad -\Gamma(\nu+1)s^{-(1+\nu)}\left\{\log s + \frac{\Gamma'(1+\nu)}{\Gamma(1+\nu)}\right\}.$

11. $B(z, \zeta) = -\frac{1}{4}\mathrm{e}^{-i\pi(z+\zeta)} \operatorname{cosec} \pi z \operatorname{cosec} \pi\zeta \int_C t^{z-1}(1-t)^{\zeta-1}\,\mathrm{d}t.$

CHAPTER 6

The Hypergeometric Equation

6.1 The differential equation

It is known [18] that the Papperitz equation

$$w'' + \left(\frac{1-\alpha-\alpha'}{z-\xi} + \frac{1-\beta-\beta'}{z-\eta} + \frac{1-\gamma-\gamma'}{z-\zeta}\right)w' + \left\{\frac{\alpha\alpha'(\xi-\eta)(\xi-\zeta)}{z-\xi} + \frac{\beta\beta'(\eta-\zeta)(\eta-\xi)}{z-\eta} + \frac{\gamma\gamma'(\zeta-\xi)(\zeta-\eta)}{z-\zeta}\right\} \times \frac{w}{(z-\xi)(z-\eta)(z-\zeta)} = 0 \quad (6.1.1)$$

is the most general linear differential equation of the second order whose only singular points are at ξ, η and ζ and are all regular. The exponents at these points are α, α'; β, β'; and γ, γ', respectively. Since the point at infinity is an ordinary point, it follows that

$$\alpha + \alpha' + \beta + \beta' + \gamma + \gamma' = 1. \quad (6.1.2)$$

The most important properties of equation (6.1.1) are:

(1) A bilinear transformation preserves the form of the Papperitz equation, transforming the position of the singular points, but preserving the exponents.

(2) A transformation of the form

$$w(z) = (z-\xi)^{-\lambda}(z-\eta)^{-\mu}(z-\zeta)^{-\nu}v(z),$$

where $\lambda + \mu + \nu = 0$, also preserves the form of the Papperitz equation, leaving the singular points unchanged, but increasing the associated exponents by λ, μ and ν respectively.

A combination of these two transformations can therefore transform the singular points to 0, 1 and ∞, and reduce one of the exponents at both 0 and 1 to zero.

If we write, as is conventional,

$$a = \alpha + \beta + \gamma, \quad b = \alpha + \beta' + \gamma, \quad c = 1 + \alpha - \alpha',$$

equation (6.1.1) becomes

$$z(1 - z)w'' + \{c - (a + b + 1)z\}w' - abw = 0, \qquad (6.1.3)$$

which is known as the hypergeometric equation.

It follows that the solution of any ordinary linear differential equation with just three singular points, all regular, can be expressed in terms of a solution of (6.1.3).

The differences of the exponents at the three singular points are respectively $1 - c$, $a + b - c$, and $a - b$, and so if any one of these is an integer we may have logarithmic terms in the general solution of the equation.

It was pointed out by Kummer that (6.1.1) was unchanged if the points ξ, η and ζ were permuted, or if either α and α' or β and β' were permuted. Thus from any one solution, 23 others may be obtained. Since two at most may be independent, a number of relations exists between them. A complete list of 24 related solutions will be found in [18].

6.2 The fundamental solution

The method of section 1.3 applied to the hypergeometric equation in the neighbourhood of the origin gives the exponents $\alpha = 0$ and $\alpha = 1 - c$ which we have already noted, and the recurrence relation for the coefficients

$$(\alpha + s)(\alpha + c - 1 + s)C_s = (\alpha + a + s - 1)(\alpha + b + s - 1)C_{s-1}. \qquad (6.2.1)$$

It follows that if c is not a negative integer we can construct a solution from the exponents $\alpha = 0$. This is

$$C_0\left\{1 + \frac{a.b}{c.1!}z + \frac{a(a+1).b(b+1)}{c(c+1)2!}z^2 + \frac{a(a+1)(a+2).b(b+1)(b+2)}{c(c+1)(c+2)3!}z^3 + \ldots\right\}. \qquad (6.2.2)$$

The function represented in the neighbourhood of the origin by the series in brackets is denoted by $F(a, b; c; z)$, and we call this the hypergeometric function. It is clear from (6.2.2) that we may with greater convenience write

$$\frac{\Gamma(a)\Gamma(b)}{\Gamma(c)}F(a, b; c; z) = \sum_{s=0}^{\infty} \frac{\Gamma(a+s)\,\Gamma(b+s)}{\Gamma(c+s)s!}z^s. \qquad (6.2.3)$$

The ratio of the coefficients of the powers of z^{s+1} and z^s in the series is

$$\frac{(a+s)(b+s)}{(c+s)(1+s)}, \tag{6.2.4}$$

which tends to 1 uniformly as $s \to \infty$, regardless of the values of a, b and c. Hence the series (6.2.3) has a radius of convergence of unity in every case.

If we assume that s is greater than the largest of $|a|$, $|b|$, $|c|$ and 1, we may expand (6.2.4) in inverse powers of s, and then the ratio takes the form

$$1 - \frac{(1+c-a-b)}{s} + O\left(\frac{1}{s^2}\right).$$

It follows from the Gauss test that the series will converge absolutely at every point on its circle of convergence if

$$\mathcal{R}(c-a-b) > 0.$$

We may also deduce that since the radius of convergence of (6.2.3) is unity, the function $F(a, b; c; z)$ has a singularity on the circle $|z| = 1$. Since this must also correspond to a singular point of the equation, it can only be at 1. The point 1 is a regular singular point of the equation, and so the singularity can be only a pole or branch point, and since we have seen above that if $\mathcal{R}(c-a-b) > 0$, the series converges at 1, it must in general be a branch point. Branch points cannot occur singly, and so it must follow that ∞ is also in general a branch point of the function.

Simple algebraic manipulation of the series for the hypergeometric equation may be used to derive recurrence relations between contiguous functions. Typical relations are

$$cF(a, b+1; c; z) = cF(a, b; c; z) + azF(a+1, b+1; c+1; z) \tag{6.2.5}$$

and

$$c\frac{\mathrm{d}}{\mathrm{d}z}F(a, b; c; z) = abF(a+1, b+1; c+1; z). \tag{6.2.6}$$

6.3 The complete solution near the origin

If the constant c is not a positive integer, we may construct a second solution from (6.2.1) using the exponent $\alpha' = 1 - c$. This solution may be written

$$z^{1-c}F(1+a-c, 1+b-c; 2-c; z). \tag{6.3.1}$$

Provided that c is not an integer, positive or negative, the general solution of the hypergeometric equation near $z = 0$ can be written as a linear combination of (6.2.3) and (6.3.1). If c is a negative integer, (6.3.1) is still a valid solution of the equation, but (6.2.3) must be replaced by a solution involving logarithmic terms, while if c is a positive integer, (6.2.3) remains valid but (6.3.1) must be so replaced.

The hypergeometric series (6.2.2) is uniformly convergent in $|z| \leqslant 1 - \delta$ for $0 < \delta < 1$. Considered as a function of c when z is within this closed disc, the function has the singularities of its individual terms. Furthermore, these singularities are simple poles at the negative integers. It also follows that we may compute the residue of $F(a, b; c; z)$ at one of these poles by summing the residues of the individual terms. Thus in particular the residue at $c = -m$ is

$$\begin{aligned}
&\lim_{c+m\to 0} (c + m)F(a, b; c; z) \\
&= \lim_{c+m\to 0} (c + m)\left\{1 + \sum_{s=1}^{\infty} \frac{a(a+1)\ldots(a+s-1)b(b+1)\ldots(b+s-1)}{c(c+1)\ldots(c+s-1)s!} z^s\right\} \\
&= (-1)^m \sum_{s=m+1}^{\infty} \frac{a(a+1)\ldots(a+s-1)b(b+1)\ldots(b+s-1)}{m!\,\{(s-m-1)!\}s!} z^s \\
&= (-1)^m \frac{\Gamma(a+m+1)\,\Gamma(b+m+1)}{\Gamma(a)\,\Gamma(b)m!} z^{m+1} \\
&\qquad \times F(a+m+1, b+m+1; 2+m; z). \qquad (6.3.2)
\end{aligned}$$

It will be seen that (6.3.2) is just a multiple of (6.3.1) with $c = -m$.

6.4 Special hypergeometric functions

A number of elementary functions can be expressed as hypergeometric functions. From section 6.2 we expect that the functions so expressed are those whose Taylor expansions have a finite radius of convergence.

We list four such functions below, representing some typical results.

(a) If $|z| < 1$, then

$$(1 - z)^{-a} = \frac{1}{\Gamma(a)} \sum_{s=0}^{\infty} \frac{\Gamma(a+s)}{s!} z^s = F(a, b; b; z). \qquad (6.4.1)$$

(b) If $|z| < 1$,

$$\log(1 + z) = z \sum_{s=0}^{\infty} (-1)^s \frac{\Gamma(s+1)}{\Gamma(s+2)} z^s = zF(1, 1; 2; -z). \qquad (6.4.2)$$

(c) If $|z| < 1$, the principal value of the function

$$\begin{aligned}
\sin^{-1} z &= \int_0^z (1 - t^2)^{-1/2}\, dt = \frac{1}{\Gamma(\frac{1}{2})} \sum_{s=0}^{\infty} \frac{\Gamma(s+\frac{1}{2})}{s!} \int_0^z t^{2s}\, dt \\
&= \frac{z}{2\Gamma(\frac{1}{2})} \sum_{s=0}^{\infty} \frac{\{\Gamma(s+\frac{1}{2})\}^2}{\Gamma(s+\frac{3}{2})s!} z^{2s} = zF(\tfrac{1}{2}, \tfrac{1}{2}; \tfrac{3}{2}; z^2). \qquad (6.4.3)
\end{aligned}$$

(d) If we assume that $|k| < 1$, the complete elliptic integral of the second kind with modulus k is defined by

$$K = \int_0^{\pi/2} (1 - k^2 \sin^2 \phi)^{-1/2}\, d\phi = \frac{1}{\Gamma(\frac{1}{2})} \sum_{s=0}^{\infty} \frac{\Gamma(s+\frac{1}{2})}{s!} k^{2s} \int_0^{\pi/2} \sin^{2s}\phi\, d\phi$$

$$= \frac{1}{2\Gamma(\frac{1}{2})} \sum_{s=0}^{\infty} \frac{\Gamma(s+\frac{1}{2}) B(s+\frac{1}{2}, \frac{1}{2})}{s!} k^{2s} = \frac{\pi}{2} F(\tfrac{1}{2}, \tfrac{1}{2}; 1; k^2). \qquad (6.4.4)$$

6.5 Alternative solutions of the hypergeometric equation

In sections 6.2 and 6.3 we have examined the solution of the equation in the neighbourhood of 0. Since 1 and ∞ are also regular singular points of the equation, we wish to find the form of the solution of the equation in other neighbourhoods.

The substitution $\zeta = 1 - z$ transforms (6.1.3) into

$$\zeta(1-\zeta)\frac{d^2w}{d\zeta^2} + \{a + b + 1 - c - (a + b + 1)\zeta\}\frac{dw}{d\zeta} - abw = 0. \qquad (6.5.1)$$

If we compare this with (6.1.3), we see that two independent solutions are

$$F(a, b; a + b + 1 - c; 1 - z) \qquad (6.5.2)$$

and

$$(1 - z)^{c-a-b} F(c - b, c - a; c + 1 - a - b; 1 - z) \qquad (6.5.3)$$

provided that $a - b - c$ is not a negative or a positive integer respectively.

The substitution $\zeta = 1/z$ transforms (6.1.3) into

$$\zeta^2(1-\zeta)\frac{d^2w}{d\zeta^2} + \zeta\{1 - a - b - (2 - c)\zeta\}\frac{dw}{d\zeta} + abw = 0. \qquad (6.5.4)$$

Then the further substitution $w(\zeta) = \zeta^\lambda v(\zeta)$ gives (6.5.4) the form

$$\zeta^2(1-\zeta)\frac{d^2v}{d\zeta^2} + \zeta\{2\lambda + 1 - a - b - (2\lambda + 2 - c)\zeta\}\frac{dv}{d\zeta}$$
$$+ \{\lambda^2 - (a + b)\lambda + ab - \lambda(\lambda + 1 - c)\zeta\}v = 0. \qquad (6.5.5)$$

If we choose $\lambda = a$ or $\lambda = b$, we can divide the equation through by ζ and reduce it to the usual form of a hypergeometric equation. If, for example, we take $\lambda = a$, it becomes

$$\zeta(1-\zeta)\frac{d^2v}{d\zeta^2} + \{a - b + 1 - (2a - c + 2)\zeta\}\frac{dv}{d\zeta}$$
$$- a(a + 1 - c)v = 0. \qquad (6.5.6)$$

Thus it follows that two solutions of the original equation are

$$z^{-a}F\left(a, a - c + 1; a - b + 1; \frac{1}{z}\right) \tag{6.5.7}$$

and

$$z^{-b}F\left(b, b - c + 1; b - a + 1; \frac{1}{z}\right) \tag{6.5.8}$$

if $a - b$ is not a negative or a positive integer, respectively.

We may extend these ideas. Let us make the bilinear substitution $\zeta = z/(z - 1)$ in (6.1.3) and so derive the equation

$$\zeta(\zeta - 1)^2\frac{\mathrm{d}^2w}{\mathrm{d}\zeta^2} + (\zeta - 1)\{(c + 1 - a - b)\zeta - c\}\frac{\mathrm{d}w}{\mathrm{d}\zeta} + abw = 0. \tag{6.5.9}$$

Following the ideas used before, put $w(\zeta) = (\zeta - 1)^a v(\zeta)$ to obtain

$$\zeta(1 - \zeta)\frac{\mathrm{d}^2v}{\mathrm{d}\zeta^2} + \{c - (c + a - b + 1)\zeta\}\frac{\mathrm{d}v}{\mathrm{d}\zeta} - a(c - b)v = 0. \tag{6.5.10}$$

and so another solution of the original equation is

$$(z - 1)^{-a}F\left\{a, c - b; c; \frac{z}{z - 1}\right\}. \tag{6.5.11}$$

We see that every one of the finite group of transformations $\zeta = z$, $1/z$, $z - 1$, $1/(z - 1)$, $z/(z - 1)$, $(z - 1)/z$, leads to an equation whose solutions can be expressed as hypergeometric functions. It is of interest to note that the region of validity of the corresponding series $|\zeta| < 1$ leads to six regions of the complex plane. Each region is the complement of a second, and has a finite intersection with the other four.

We have already seen that the only finite singularity of $F(a, b; c; z)$ is at 1 and it is in general a branch point. We are therefore able to continue the function outside the circle $|z| < 1$. This can formally be done by expressing it as a linear combination of independent solutions of an equation derived by making one of the substitutions above. It is a more difficult problem to determine the coefficients in this linear combination, and we may most easily accomplish this by using an integral representation, as we show below.

6.6 A representation by an Euler-type integral

We have shown in section 2.10 that we can find solutions of the hypergeometric equation of the form

$$\int_C t^{b-1}(t - 1)^{c-b-1}(zt - 1)^{-a}\,\mathrm{d}t \tag{6.6.1}$$

where C is a double-dumb-bell contour, of the type described in Chapter 2. In particular, if $\mathcal{R}(c) > \mathcal{R}(b) > 0$, we can reduce the contour integral to a real integral

$$\int_0^1 t^{b-1}(1-t)^{c-b-1}(1-zt)^{-a}\,dt. \tag{6.6.2}$$

If we assume $|z| < 1$, then in $0 \leqslant t \leqslant 1$ we can expand $(1-zt)^{-a}$ in a uniformly convergent series, and so we can express (6.6.2) in the form

$$\frac{1}{\Gamma(a)}\sum_{s=0}^{\infty}\frac{\Gamma(a+s)}{s!}z^s\int_0^1 t^{b+s-1}(1-t)^{c-b-1}\,dt$$
$$=\frac{1}{\Gamma(a)}\sum_{s=0}^{\infty}\frac{\Gamma(a+s)}{s!}B(b+s,c-b)z^s.$$

Hence if $\mathcal{R}(c) > \mathcal{R}(b) > 0$,

$$F(a,b;c;z)=\frac{\Gamma(c)}{\Gamma(c-b)\,\Gamma(b)}\int_0^1 t^{b-1}(1-t)^{c-b-1}(1-zt)^{-a}dt \tag{6.6.3}$$

and the right-hand side is a regular function of z, in the whole complex plane cut from 1 to ∞ along the positive axis. This, then, will enable us to perform the analytic continuation to the rest of the plane, under the restrictions made above.

We may deduce immediately a relation of the form suggested in the previous section by substituting $t = 1 - \tau$ into the integral:

$$F(a,b,c,z)=\frac{\Gamma(c)}{\Gamma(c-b)\,\Gamma(b)}\int_0^1 \tau^{c-b-1}(1-\tau)^{b-1}(1-z+z\tau)^{-a}\,d\tau$$
$$=\frac{\Gamma(c)(1-z)^{-a}}{\Gamma(c-a)\,\Gamma(b)}\int_0^1 \tau^{c-b-1}(1-\tau)^{b-1}\left(1-\frac{z\tau}{z-1}\right)^{-a}d\tau$$
$$=(1-z)^{-a}F\left(a,c-b;c;\frac{z}{z-1}\right). \tag{6.6.4}$$

Another simple deduction from (6.6.3) is that if in addition $\mathcal{R}(c-a-b) > 0$, then

$$F(a,b;c;1)=\frac{\Gamma(c)}{\Gamma(c-b)\,\Gamma(b)}B(b,c-b-a)=\frac{\Gamma(c)\,\Gamma(c-a-b)}{\Gamma(c-b)\,\Gamma(c-a)}. \tag{6.6.5}$$

We notice that we can remove the restrictions $\mathcal{R}(c) > \mathcal{R}(b) > 0$, if we generalize (6.6.3) by the use of an integral like (6.6.1). The resulting integral is then valid for all c, and in particular it is valid even when c is a negative integer. It is clear that as c approaches a negative integer, the limiting form of the integral is a regular function of z, and so involves the residue of $F(a, b; c; z)$, considered as a function of c. We have shown in section 6.3, however, that this is proportional to the second solution.

Any other contour which is acceptable in the sense of section 2.10 will give in the general case a linear combination of the two fundamental solutions, and so in the limit, when c tends to an integer, gives a logarithmic term.

6.7 The Barnes integral

A second form of integral representation is given by the Barnes integral. This is of greater practical value than that given in the previous section.

Suppose that $\mathscr{R}(a) > 0, \mathscr{R}(b) > 0$, and consider

$$\frac{1}{2\pi i}\int_C \frac{\Gamma(a+\zeta)\,\Gamma(b+\zeta)\,\Gamma(-\zeta)(-z)^\zeta}{\Gamma(c+\zeta)}\,d\zeta \tag{6.7.1}$$

where C is a contour joining $-i\infty$ to $i\infty$ which passes to the right of the poles of $\Gamma(a+\zeta)$ and $\Gamma(b+\zeta)$ and to the left of the poles of $\Gamma(-\zeta)$. Since the integrand is implicitly a function of $\log(-z)$, (6.7.1) will represent a function regular in $|\arg(-z)| \leqslant \pi - \delta$, for every $0 < \delta < \pi$.

Let us suppose that z lies within this sector for some $0 < \delta < \pi$, and suppose also that $|z| < 1$. We consider an integral of the form (6.7.1) but in which the contour C has been replaced by C_*. The closed contour C_* consists of the imaginary ζ-axis for $|\zeta| \leqslant N + \frac{1}{2}$, indented to the left of the origin, and a semicircle of radius $N + \frac{1}{2}$ with centre at $\zeta = 0$ and to the right of the imaginary axis. We will assume that N is a positive integer.

From (5.4.6), we may replace $\Gamma(-\zeta)$ in the integrand by $-\{\pi/\Gamma(\zeta+1)\}$ cosec $\pi\zeta$. Now put $\zeta = (N+\frac{1}{2})e^{i\theta}$ on the semicircle and we see that the parts of the integrand can be separated, and their orders of magnitude assessed. We use Stirling's approximation to see that

$$\left|\frac{\Gamma(a+\zeta)\,\Gamma(b+\zeta)}{\Gamma(c+\zeta)\,\Gamma(1+\zeta)}\right| = O(N^{a+b-c-1}).$$

Also on the same semicircle

$$|\operatorname{cosec}\zeta\pi| = O[\exp\{-(N+\tfrac{1}{2})\pi\,|\sin\theta|\}]$$

and

$$|(-z)^\zeta| = O[\exp\{(N+\tfrac{1}{2})\cos\theta\log|z| + (N+\tfrac{1}{2})(\pi-\delta)|\sin\theta|\}].$$

Thus the contribution from the semicircle is exponentially small provided $\log|z| < 0$. This, however, is an immediate consequence of the assumption that $|z| < 1$.

If we let $N \to \infty$, the integrals round C and C_* are equal and their value can be obtained from the residue theorem, namely

$$\sum_{s=0}^{\infty} \frac{\Gamma(a+s)\,\Gamma(b+s)}{\Gamma(c+s)\,\Gamma(1+s)}(-1)^s\,|\,(-z)^s\,| = \frac{\Gamma(a)\,\Gamma(b)}{\Gamma(c)} F(a, b; c; z). \quad (6.7.2)$$

We may similarly show that if $|\,z\,| > 1$ we may close the contour C by a semicircle to the left. The enclosed poles will now be those at $-a-s$, and $-b-s$ for $s = 0, 1, 2, \ldots$ In this case, therefore, the residue theorem will give for the value of the integral

$$\sum_{s=0}^{\infty} (-1)^s \left\{ \frac{\Gamma(b-a-s)\,\Gamma(a+s)}{\Gamma(c-a-s)\,\Gamma(1+s)}(-z)^{-a-s} + \frac{\Gamma(a-b-s)\,\Gamma(b+s)}{\Gamma(c-b-s)\,\Gamma(1+s)}(-z)^{-b-s} \right\}. \quad (6.7.3)$$

To express these terms in the familiar form of the hypergeometric series we use relations of the form

$$\begin{aligned} \frac{\Gamma(b-a-s)}{\Gamma(c-a-s)} &= \frac{\Gamma(b-a)}{\Gamma(c-a)} \frac{(c-a-1)(c-a-2)\ldots(c-a-s)}{(b-a-1)(b-a-2)\ldots(b-a-s)} \\ &= \frac{\Gamma(b-a)}{\Gamma(c-a)} \frac{(1+a-c)(2+a-c)\ldots(s+a-c)}{(1+a-b)(2+a-b)\ldots(s+a-b)} \\ &= \frac{\Gamma(b-a)\,\Gamma(1+a-b)\,\Gamma(a+1-c+s)}{\Gamma(c-a)\,\Gamma(1+a-c)\,\Gamma(a+1-b+s)}. \end{aligned}$$

Then we deduce that

$$\begin{aligned} F(a, b; c; z) &= \frac{\Gamma(c)\,\Gamma(b-a)}{\Gamma(b)\,\Gamma(c-a)}(-z)^{-a} F(a, a-c+1; a-b+1; z^{-1}) \\ &\quad + \frac{\Gamma(c)\,\Gamma(a-b)}{\Gamma(a)\,\Gamma(c-b)}(-z)^{-b} F(b, b-c+1; b-a+1; z^{-1}). \quad (6.7.4) \end{aligned}$$

We may continue this result into the region $\mathcal{R}(a) < 0, \mathcal{R}(b) < 0$ if we use the recurrence relation (6.2.5).

6.8 Generalized hypergeometric functions

A direct generalization of the hypergeometric function (see [2]) is

$$\begin{aligned} &F(a_1, a_2, \ldots a_k; c_1, c_2, \ldots c_j; z) \\ &\quad = \frac{\Gamma(c_1)\,\Gamma(c_2)\ldots\Gamma(c_j)}{\Gamma(a_1)\,\Gamma(a_2)\ldots\Gamma(a_k)} \sum_{s=0}^{\infty} \frac{\Gamma(a_1+s)\,\Gamma(a_2+s)\ldots\Gamma(a_k+s)}{\Gamma(c_1+s)\,\Gamma(c_2+s)\ldots\Gamma(c_j+s)s!} z^s. \quad (6.8.1) \end{aligned}$$

This function is commonly denoted by

$$ {}_kF_j(a; c; z), \quad (6.8.2)$$

where the first suffix k gives the number of members of the set (a) and the second j the number of members of the set (c).

In this notation the ordinary hypergeometric function is

$$ {}_2F_1(a; c; z). \tag{6.8.3}$$

If we consider the ratio of successive coefficients in the series (6.8.1) it is clear that, if $k > j + 1$ the series can converge only at $z = 0$, if $k = j + 1$ it must have unit radius of convergence, and if $k \leqslant j$ the series represents an entire function of z.

Exercises

1. If $F(a, b; c; z)$ is defined by the series in $|z| < 1$, show that

$$z(1 - z)\frac{d^2F}{dz^2} = b(a + b + 1)F(a, b + 1; c; z) - abF(a + 1, b + 1; c + 1; z) - b(b + 1)F(a, b; c; z).$$

Hence verify that $F(a, b; c; z)$ is a solution of the hypergeometric equation

$$z(1 - z)w'' + \{c - (a + b + 1)z\}w' - abw = 0.$$

[*Hint*: show that $(b + 1)F(a + 1,\ b + 2;\ c + 1;\ z) + (c - b - 1) \times F(a + 1, b + 1; c + 1; z) = cF(a + 1, b + 1; c; z)$.]

2. Verify the following relations:

(i) $\tanh^{-1} z = zF(\frac{1}{2}, 1; \frac{3}{2}; z^2), \qquad |z| < 1$

(ii) $E = \displaystyle\int_0^{\pi/2} \{1 - k^2 \sin^2 \phi\}^{1/2}\, d\phi = \frac{\pi}{2}F(-\tfrac{1}{2}, \tfrac{1}{2}; 1; k^2), \qquad |k| < 1.$

3. If $z = \sin^2 \zeta/2$, $T_\nu(z) = \cos \nu\zeta$, show that

$$z(1 - z)\frac{d^2T_\nu}{dz^2} + (\tfrac{1}{2} - z)\frac{dT_\nu}{dz} + \nu^2 T_\nu = 0.$$

Hence or otherwise verify that

$$T_\nu(z) = F(\nu, -\nu; \tfrac{1}{2}; z).$$

4. If $\mathscr{R}(c) > \mathscr{R}(b) > 0$, show that

$$(1 - z)^{a+b-c}\int_0^1 t^{b-1}(1 - t)^{c-b-1}(1 - tz)^{-a}\, dt = \int_0^{1/(1-z)} \{1 - (1 - z)\eta\}^{b-1}\eta^{c-b-1}(1 + \eta z)^{-a}\, d\eta,$$

by writing $1 - t = (1 - z)\eta$.

Hence deduce that if $\mathscr{R}(c) > \mathscr{R}(b) > 0$, but $\mathscr{R}(c - a - b) < 0$, then

$$\lim_{z\to 1-0} (1 - z)^{a+b-c}F(a, b; c; z) = \frac{\Gamma(a + b - c)\,\Gamma(c)}{\Gamma(a)\,\Gamma(b)}.$$

5. Use the result of Exercise 4 to show that

$$F(a, b; c; z) = \frac{\Gamma(c)\,\Gamma(c-a-b)}{\Gamma(c-b)\,\Gamma(c-a)} F(a, b; a+b+1-c; 1-z)$$
$$+ \frac{\Gamma(c)\,\Gamma(a+b-c)}{\Gamma(a)\,\Gamma(b)} (1-z)^{c-a-b} F(c-a, b-a;$$
$$c+1-a-b; 1-z).$$

[*Hint*: if the constants on the right would be valid for some range of the constants a, b and c, the relation will hold for all other values for which it is meaningful, by the principle of analytic continuation.]

6. The two-dimensional steady flow of an inviscid compressible fluid satisfies the hodograph equations, which are

$$\frac{\partial \phi}{\partial \tau} = P(\tau)\frac{\partial \psi}{\partial \theta}, \qquad \frac{\partial \phi}{\partial \theta} = Q(\tau)\frac{\partial \psi}{\partial \tau}.$$

In these equations $\phi(\tau, \theta)$ is a velocity potential, $\psi(\tau, \theta)$ is a stream function, θ is the angle the velocity makes with a fixed direction and $\tau = q^2/q_{\max}{}^2$, where q is the speed of the flow and $q_{\max}$ is the greatest speed that can be obtained. For a homentropic gas

$$2\tau(1-\tau)^{\gamma/(\gamma-1)}P(\tau) = \frac{\gamma+1}{\gamma-1}\tau - 1, \quad (1-\tau)^{1/(\gamma-1)}Q(\tau) = 2\tau,$$

where γ is the ratio of specific heats and $1 < \gamma < 2$. Show that a solution of these equations for the stream function may have the form

$$\psi(\tau, \theta) = \tau^{\nu/2} F(a_\nu, b_\nu; 1+\nu; \tau) e^{-i\nu\theta}$$

where $a_\nu + b_\nu = \nu - 1/(\gamma - 1)$ and $2(\gamma - 1)a_\nu b_\nu = -\nu(\nu + 1)$.

CHAPTER 7

The Confluent Hypergeometric Function

7.1 The confluent function as a limit

From the previous chapter we see that if c is not a negative integer,

$$F\left(a, b; c; \frac{z}{b}\right) = \frac{\Gamma(c)}{\Gamma(a)}\left\{\frac{\Gamma(a)}{\Gamma(c)} + \frac{\Gamma(a+1)\,bz}{\Gamma(c+1)\,b.1} + \frac{\Gamma(a+2)}{\Gamma(c+2)}\frac{b(b+1)z^2}{b^2.2!} + \ldots \right.$$
$$\left. + \frac{\Gamma(a+n)}{\Gamma(c+n)}.\frac{b(b+1)\ldots(b+n-1)}{b^n}\frac{z^n}{n!} + \ldots\right\}, \quad (7.1.1)$$

and this series clearly has $|b|$ as its radius of convergence.

Choose an arbitrarily large number R and suppose $|z| \leqslant R$. Then the series (7.1.1) is uniformly convergent in z for all b with $|b| > 2R$.

Now for fixed n

$$\lim_{|b|\to\infty} \frac{b(b+1)\ldots(b+n-1)}{b^n} = \lim_{|b|\to\infty}\left(1+\frac{1}{b}\right)\left(1+\frac{2}{b}\right)\ldots$$
$$\times\left(1+\frac{n-1}{b}\right) = 1. \quad (7.1.2)$$

Since the series (7.1.1) is uniformly convergent, the limit of the sum is equal to the sum of the limits. Hence

$$\lim_{b|\to\infty} F\left(a, b; c; \frac{z}{b}\right) = \frac{\Gamma(c)}{\Gamma(a)}\sum_{n=0}^{\infty}\frac{\Gamma(a+n)z^n}{\Gamma(c+n)n!}\lim_{|b|\to\infty}\left(1+\frac{1}{b}\right)\left(1+\frac{2}{b}\right)\ldots$$
$$\times\left(1+\frac{n-1}{b}\right)$$

$$= \frac{\Gamma(c)}{\Gamma(a)}\sum_{n=0}^{\infty}\frac{\Gamma(a+n)z^n}{\Gamma(c+n)n!}. \quad (7.1.3)$$

In terms of the notation of section 6.8

$$\lim_{|b|\to\infty} F\left(a, b; c; \frac{z}{b}\right) = {}_1F_1(a; c; z). \tag{7.1.4}$$

Since the choice of R was arbitrary the expansion (7.1.3) must represent an entire function. This is obvious from the expansion itself, and is also in accord with the remark made in section 6.8.

If we return once again to the limiting process, it is clear that at any z with $|z| \leqslant R$, the first and second derivatives of $F(a, b; c; z/b)$ are uniformly bounded for all b with $|b| \geqslant 2R$. We can therefore apply the same limiting process to the differential equation. Now $F(a, b; c; z/b)$ satisfies the equation

$$z(b - z)w'' + \{bc - (a - b - 1)z\}w' - abw = 0. \tag{7.1.5}$$

When we divide (7.1.5) by b and let b tend to infinity, it follows that ${}_1F_1(a; c; z)$ satisfies the differential equation

$$zw'' + (c - z)w' - aw = 0. \tag{7.1.6}$$

The differential equation (7.1.6) has as its only singular points a regular point at 0, and an irregular point at ∞. From the point of view of the limiting process described above, the irregular point at ∞ has been formed by the confluence of the two regular points of equation (7.1.5) at b and ∞; hence the name 'confluent' hypergeometric function applied to ${}_1F_1(a; c; z)$.

The expansion (7.1.3) is subject to the same restriction as that for the expansion of $F(a, b; c; z)$, namely that c is not a negative integer. Similarly, we see that if c is not a positive integer a second independent solution of (7.1.6), valid near 0, is

$$z^{1-c}{}_1F_1(1 + a - c; 2 - c; z). \tag{7.1.7}$$

Like the ordinary hypergeometric function, the confluent function (7.1.3) considered as a function of c, has simple poles at the negative integers. For fixed z, the residue at $-m$ is proportional to

$$z^{1+m}{}_1F_1(1 + a + m; 2 + m; z).$$

This is the second solution with the substitution $c = -m$.

7.2 Special confluent hypergeometric functions

In the previous chapter we saw that some familiar functions expansible as series with finite radii of convergence were hypergeometric functions; others with infinite radii of convergence are confluent hypergeometric functions. Illustrations are:

(a) The exponential function

$$e^z = \sum_{n=0}^{\infty} \frac{z^n}{n!} = {}_1F_1(a; a; z); \tag{7.2.1}$$

(b) For $\mathscr{R}(\alpha) > 0$, the incomplete gamma function

$$\int_0^z e^{-t} t^{\alpha-1}\, dt = \sum_{n=0}^{\infty} \frac{(-1)^n}{n!} \int_0^z t^{n+\alpha-1}\, dt = z^\alpha \sum_{n=0}^{\infty} \frac{(-1)^n}{n+\alpha} \frac{z^n}{n!}$$

$$= z^\alpha \sum_{n=0}^{\infty} \frac{\Gamma(n+\alpha)}{\Gamma(n+\alpha+1)} \frac{(-z)^n}{n!} = \alpha^{-1} z^\alpha {}_1F_1(\alpha; \alpha+1; -z); \tag{7.2.2}$$

(c) The error function, a particular case of (b),

$$\operatorname{erf}(z) = \frac{2}{\sqrt{\pi}} \int_0^z \exp(-t^2)\, dt = \frac{2}{\sqrt{\pi}} \sum_{n=0}^{\infty} \frac{(-1)^n}{2n+1} \frac{z^{2n+1}}{n!}$$

$$= \frac{z}{\sqrt{\pi}} \sum_{n=0}^{\infty} \frac{\Gamma(n+\frac{1}{2})}{\Gamma(n+\frac{3}{2})} \frac{(-z^2)^n}{n!}$$

$$= \frac{2}{\sqrt{\pi}} z\, {}_1F_1(\tfrac{1}{2}; \tfrac{3}{2}; -z^2). \tag{7.2.3}$$

We shall show in later chapters that the Bessel, Laguerre and Hermite functions can be expressed in terms of confluent hypergeometric functions.

7.3 Integrals for the confluent function

Since the ordinary hypergeometric function is symmetric in a and b, we can use the first integral of section 6.6 (equation (6.6.3)) to give

$$F\left(a, b; c; \frac{z}{b}\right) = \frac{\Gamma(c)}{\Gamma(c-a)\,\Gamma(a)} \int_0^1 t^{a-1}(1-t)^{c-a-1}\left(1 - \frac{zt}{b}\right)^{-b} dt. \tag{7.3.1}$$

Intuitively we would suppose that in the limit $|b| \to \infty$, the last bracket would be replaced by e^{zt}. It is easy to justify this limiting process by uniform convergence, but it is easier to proceed directly.

If $\mathscr{R}(c) > \mathscr{R}(a) > 0$,

$$\int_0^1 e^{zt} t^{a-1}(1-t)^{c-a-1}\, dt = \sum_{n=0}^{\infty} \frac{z^n}{n!} \int_0^1 t^{n+a-1}(1-t)^{c-a-1}\, dt$$

$$= \sum_{n=0}^{\infty} B(n+a, c-a) \frac{z^n}{n!}$$

$$= \Gamma(c-a) \sum_{n=0}^{\infty} \frac{\Gamma(a+n)\, z^n}{\Gamma(c+n)\, n!}$$

$$= \frac{\Gamma(c-a)\,\Gamma(a)}{\Gamma(c)} {}_1F_1(a; c; z). \tag{7.3.2}$$

The term-by-term integration is justified, since e^{zt} is an entire function.

The substitution $t = 1 - \tau$ into the integral gives an interesting relation between functions:

$$\Gamma(c-a)\,\Gamma(a)\,{}_1F_1(a; c; z) = \Gamma(c)\int_0^1 e^{zt}t^{a-1}(1-t)^{c-a-1}\,dt,$$

$$= e^z\,\Gamma(c)\int_0^1 e^{-z\tau}\tau^{c-a-1}(1-\tau)^{a-1}\,d\tau,$$

$$= e^z\,\Gamma(c-a)\,\Gamma(a)\,{}_1F_1(c-a; c; -z).$$

Hence

$$ {}_1F_1(a; c; z) = e^z\,{}_1F_1(c-a; c; -z). \tag{7.3.3}$$

If $\pi \geqslant |\arg z| \geqslant \pi/2 + \delta$, the form of the integral (7.3.2) suggests that we find the asymptotic expansion by using Watson's lemma. Then (7.3.3) provides the corresponding result in the sector $|\arg z| \leqslant \pi/2 - \delta$. We postpone consideration of the asymptotic form to a later section and find it by a different approach.

We may remove the restrictions on c and a if we replace the line integral by a contour integral. For the details see section 2.5.

7.4 Another Laplace-type integral

If we solve the original equation (7.1.6) by a contour integral of the Laplace type, it is easily seen that a solution in the half plane $\mathscr{R}(z) > 0$ is

$$U(a; c; z) = \int_0^\infty e^{-zt}t^{a-1}(1+t)^{c-a-1}\,dt, \tag{7.4.1}$$

provided $\mathscr{R}(a) > 0$.

This is therefore of the form

$$A\,{}_1F_1(a; c; z) + Bz^{1-c}\,{}_1F_1(1+a-c; 2-c; z) \tag{7.4.2}$$

for some constants A and B, provided that c is not an integer, in order to make this expression meaningful. We may determine these constants by applying the same limiting process to (7.4.1) and (7.4.2) separately. Provided that this process is applicable in a whole domain of the complex c-plane, we are justified in continuing the result analytically to all points of the c-plane at which both sides are meaningful.

First suppose that not only $\mathscr{R}(a) > 0$, but also $\mathscr{R}(1-c) > 0$. Then as $z \to 0$ in a suitable cut z-plane, (7.4.2) tends uniformly to the limit A. On the other hand (7.4.1) tends uniformly to the convergent integral

$$\int_0^\infty t^{a-1}(1+t)^{c-a-1}\,dt.$$

This can be reduced to a beta function by the transformation $t = \eta/(1-\eta)$ (see Exercise 9(b), Chapter 5). Hence

$$A = \frac{\Gamma(a)\,\Gamma(1-c)}{\Gamma(1+a-c)}. \tag{7.4.3}$$

Next suppose that $-1 < \mathscr{R}(1-c) < 0$, so that from (7.4.1) and (7.4.2) we find that

$$B = \lim_{x \to +0} x^{c-1} \int_0^\infty e^{-xt} t^{a-1}(1+t)^{c-a-1}\,dt.$$

We write x instead of z in the integral to stress that the limit is to be taken as $z \to 0$ along the positive real axis. If we replace xt by η, this takes the alternative form

$$B = \lim_{x \to +0} \int_0^\infty e^{-\eta}\eta^{a-1}(x+\eta)^{c-a-1}d\eta.$$

Choose $\varepsilon > 0$, and write

$$\int_0^\infty e^{-\eta}\eta^{a-1}(x+\eta)^{c-a-1}\,d\eta = \int_0^\varepsilon e^{-\eta}\eta^{a-1}(x+\eta)^{c-a-1}\,d\eta + \int_\varepsilon^\infty e^{-\eta}\eta^{a-1}(x+\eta)^{c-a-1}\,d\eta,$$
$$= I_1(x, \varepsilon) + I_2(x, \varepsilon),$$

say. For every x in the closed interval $[0, x_0]$, where x_0 is arbitrary but finite, $I_1(x, \varepsilon)$ is of order ε as $\varepsilon \to 0$. For fixed ε as $x \to +0$, the integral $I_2(x, \varepsilon)$ differs by a term which is $O(\varepsilon)$ from $\int_0^\infty e^{-\eta}\eta^{c-2}\,d\eta$. It follows then that

$$B = \Gamma(c-1). \tag{7.4.4}$$

Finally, therefore, if $\mathscr{R}(a) > 0$, and c is not an integer,

$$U(a; c; z) = \frac{\Gamma(a)\,\Gamma(1-c)}{\Gamma(1+a-c)}\,{}_1F_1(a; c; z) + \Gamma(c-1)z^{1-c}{}_1F_1(1+a-c; 2-c; z). \tag{7.4.5}$$

7.5 Further properties of the Laplace-type integral

It is clear from (7.4.1), however, that the integral is well defined for all complex c, and so the right-hand side must also be valid even if c is an integer. It is fairly easy to see that both constituents of the right-hand side of (7.4.5) considered as functions of c separately have simple poles at the integers, and the residues are equal and opposite in sign. It follows that the limit of the right-hand side of (7.4.5) as c tends to an integer is finite, but to obtain an explicit form for it requires lengthy manipulation.

When we let c tend to a positive integer greater than or equal to unity, for example, we obtain

$$\Gamma(1+a-M)U(a; M; z)$$
$$= \sum_{s=0}^{M-2} \Gamma(1+a-M+s)\,\Gamma(M-s-1)\,\frac{z^{s+1-M}}{s!}$$
$$+ (-1)^M \left[\frac{\Gamma(a)}{\Gamma(M)} \log z\, {}_1F_1(a; M; z)\right.$$
$$\left. + \sum_{s=0}^{\infty} \frac{\Gamma(a+s)}{\Gamma(M+s)} \{\psi(a+s) - \psi(M+s) - \psi(1+s)\} \frac{z^s}{s!}\right], \quad (7.5.1)$$

where we have written $\psi(\zeta) = \Gamma'(\zeta)/\Gamma(\zeta)$.

It will also be observed that the right-hand sides of both (7.4.5) and (7.5.1) are analytic when considered as functions of a, except for simple poles at the negative integers. This is clear if we replace (7.4.1) by

$$U(a; c; z) = \operatorname{cosec} a\pi \frac{1}{2i} \int_C e^{-zt} t^{a-1} (1+t)^{c-a-1}\, dt \quad (7.5.2)$$

where C is a contour starting at ∞ on the positive real axis with $\arg t = 0$, encircling the origin, passing between 0 and -1, and returning to $+\infty$.

It appears from (7.4.5) that in general $U(a; c; z)$ is a many-valued function of z. Since ${}_1F_1(a; c; z)$ is an entire function, we see that if the argument of z is increased by 2π we derive

$$U(a; c; ze^{2\pi i}) = \frac{\Gamma(a)\,\Gamma(1-c)}{\Gamma(1+a-c)} {}_1F_1(a; c; z)$$
$$+ \Gamma(c-1) z^{1-c} e^{-2c\pi i} {}_1F_1(1+a-c; 2-c; z).$$

Eliminating ${}_1F_1(1+a-c; 2-c; z)$ by the use of (7.4.5), we have

$$U(a; c; ze^{2\pi i}) = e^{-2c\pi i} U(a; c; z)$$
$$+ \frac{\Gamma(a)\,\Gamma(1-c)}{\Gamma(1+a-c)} (1 - e^{-2c\pi i}) {}_1F_1(a; c; z). \quad (7.5.3)$$

7.6 The Barnes integral

We may apply the limiting process of section 7.1 to the Barnes integral for the hypergeometric function. We write

$$\frac{\Gamma(c)}{\Gamma(a)} F\left(a, b; c; \frac{z}{b}\right) = \frac{1}{2\pi i} \int_C \frac{\Gamma(a+s)\,\Gamma(-s)}{\Gamma(c+s)} \left\{\frac{\Gamma(b+s)}{b^s\,\Gamma(b)}\right\} (-z)^s\, ds,$$

where C is the same contour as in section 6.7. Since in that section it was assumed that $\mathscr{R}(a) > 0$, $\mathscr{R}(b) > 0$, the limit $|b| \to \infty$ must be taken within a sector $|\arg b| \leqslant \pi/2 - \delta$. For fixed s, it follows from Stirling's formula that

$$\lim_{|b| \to \infty} \frac{\Gamma(b+s)}{b^s \Gamma(b)} = 1.$$

Thus it follows that, provided we can reverse the order of the limit and the integration,

$$\frac{\Gamma(c)}{\Gamma(a)}{}_1F_1(a; c; z) = \frac{1}{2\pi i}\int_C \frac{\Gamma(a+s)\,\Gamma(-s)}{\Gamma(c+s)}(-z)^s\,ds. \qquad (7.6.1)$$

Again (7.6.1) can be established directly by the expansion of the integral, rather than by justifying the limiting process above. The expansion follows in a similar way to that of section 6.7, but its validity is more easily established than in the case of the hypergeometric function. If in this case we close the contour by a semicircle to the right of the imaginary axis of radius $N + \frac{1}{2}$, for integer N, then from Stirling's formula we show that on this semicircle

$$\left|\frac{\Gamma(a+s)}{\Gamma(c+s)\,\Gamma(1+s)}\right| \quad \text{is} \quad O(N^{a-c-1/2-\sigma}e^{\sigma})$$

where $\sigma = \mathscr{R}(s)$. Thus this modulus tends quickly to zero away from $s = \pm i(N + \frac{1}{2})$. It follows that the Barnes integral can be computed as the sum of the residues at the poles of $\Gamma(-s)$, and (7.6.1) is derived.

7.7 The Whittaker Functions

The confluent hypergeometric function ${}_1F_1(a; c; z)$ is easily defined, as has been shown in the earlier sections. It is not, however, the most convenient function for practical application, for reasons that will appear when we consider the asymptotic expansion. It is usual to express it in terms of another pair of functions, more complicated to define, but with the more fundamental asymptotic form.

We substitute $w(z) = z^{-c/2}e^{z/2}v(z)$ into equation (7.1.6) to obtain

$$v''(z) - \left\{\frac{1}{4} + \frac{2a-c}{2z} + \frac{c(c-2)}{4z^2}\right\}v(z) = 0.$$

The canonical form of this equation is obtained if we make the further substitution $2k = c - 2a$, $2m = c - 1$, when it becomes the Whittaker equation

$$v''(z) - \left\{\frac{1}{4} - \frac{k}{z} - \frac{1-4m^2}{4z^2}\right\}v(z) = 0. \qquad (7.7.1)$$

If we now use the results established in section 7.4 we see that (7.7.1) has a solution which may be written

$$\frac{z^{m+1/2}}{\Gamma(m-k+\frac{1}{2})}e^{-z/2}U(m-k-\tfrac{1}{2}; 2m+1; z). \qquad (7.7.2)$$

This solution, which is regular in the half plane $\mathscr{R}(z) > 0$, is conventionally denoted by $W_{k,m}(z)$. It was pointed out in section 7.5 that $U(a; c; z)$ was defined for all complex a, excepting only the negative integers. It is easily seen that $\{1/\Gamma(a)\}U(a; c; z)$ has a finite limit as $a \to -N$, a negative integer, and in fact this limit is a polynomial in z of order N. It follows that (7.7.2) is well defined for all complex k and m.

The equation (7.7.1) is left invariant if we simultaneously change the signs of both k and z. Thus another solution of (7.7.1) is $W_{-k,m}(-z)$. We shall show in the next section that $W_{k,m}(z) \to 0$ exponentially as $|z| \to \infty$ in the right-hand half plane. Then $W_{-k,m}(-z)$ is exponentially large and it represents an independent second solution.

If we combine (7.7.2) and (7.4.1) and make the substitution $\eta = zt$, then in the half plane $\mathscr{R}(z) > 0$

$$W_{k,m}(z) = \frac{z^k e^{-z/2}}{\Gamma(m-k+\frac{1}{2})} \int_0^\infty e^{-\eta} \eta^{m-k-1/2} \left(1 + \frac{\eta}{z}\right)^{m+k-1/2} d\eta. \quad (7.7.3)$$

Now the right-hand side is regular in the whole z-plane cut along the negative real axis. We have therefore an analytic continuation of $W_{k,m}(z)$ into the larger domain.

The integral (7.7.3) still converges if the line of integration is distorted off the real axis to the line $\arg \eta = \alpha$ provided that $|\alpha| < \pi/2$, so that the integrand is exponentially small for large $|\eta|$.

When the line of integration is taken as $\arg \eta = \alpha$, the integral defines a function regular in $\alpha - \pi < \arg z < \pi + \alpha$. Thus by varying α we can continue the definition of $W_{k,m}(z)$ into the region

$$-\frac{3\pi}{2} < \arg z < \frac{3\pi}{2}.$$

Despite this, $W_{k,m}(z)$ is in general a many-valued function and we expect that in almost all cases

$$\lim_{\theta \to \pi} \{W_{k,m}(r\, e^{i\theta}) - W_{k,m}(r\, e^{-i\theta})\} \neq 0,$$

for fixed r.

7.8 The asymptotic expansion of the Whittaker functions

The function $(1 + t)^{m+k-1/2}$ is regular in the whole complex t-plane cut from $-\infty$ to -1 along the negative real axis. For $|t| < 1$ we have the Taylor series

$$(1 + t)^{m+k-1/2} = \Gamma(m + k + \tfrac{1}{2}) \sum_{s=0}^{\infty} \frac{t^s}{\Gamma(m + k + \frac{1}{2} - s)s!},$$

while if $|t|$ is very large (in the cut plane)

$$(1 + t)^{m+k-1/2} = t^{m+k-1/2}\left\{1 + O\left(\frac{1}{|t|}\right)\right\}.$$

Choose $N > \mathscr{R}(m + k - \frac{1}{2})$ and consider the function

$$t^{-N}\left\{(1 + t)^{m+k-1/2} - \Gamma(m + k + \tfrac{1}{2})\sum_{s=0}^{N-1}\frac{t^s}{\Gamma(m + k + \frac{1}{2} - s)s!}\right\}.$$

It is clear from the Taylor expansion that this function is bounded near 0, and is also bounded as $|t| \to \infty$ in the cut plane. It is therefore regular and bounded at every point of the cut t-plane. Hence we can find a constant K_N, which must depend on N, so that

$$\left|(1 + t)^{m+k-1/2} - \Gamma(m + k + \tfrac{1}{2})\sum_{s=0}^{N-1}\frac{t^s}{\Gamma(m + k + \frac{1}{2} - s)s!}\right| \leqslant K_N|t|^N.$$

Now we apply this result to the right-hand side of (7.7.3) and derive

$$\left|\int_0^\infty e^{-\eta}\eta^{m-k+1/2}\left(1 + \frac{\eta}{z}\right)^{m+k-1/2} d\eta - \Gamma(m + k + \tfrac{1}{2})\right.$$
$$\left.\times \sum_{s=0}^{N-1}\frac{z^{-s}}{\Gamma(m + k + \frac{1}{2} - s)s!}\int_0^\infty e^{-\eta}\eta^{m+k+s-1/2}\, d\eta\right|$$
$$\leqslant \frac{K_N}{|z|^N}\int_0^\infty e^{-\eta}\eta^{m-k+N-1/2}\, d\eta.$$

Then

$$\left|e^{z/2}z^{-k}W_{k,m}(z) - \frac{\Gamma(m + k + \frac{1}{2})}{\Gamma(m - k + \frac{1}{2})}\sum_{s=0}^{N-1}\frac{\Gamma(m - k + s + \frac{1}{2})}{\Gamma(m + k - s + \frac{1}{2})s!z^s}\right|$$
$$\leqslant K_N'|z|^{-N} \quad (7.8.1)$$

for a new constant K_N'. This establishes that the series is an asymptotic expansion for $W_{k,m}(z)$, valid in $|\arg z| \leqslant \pi - \delta$ for $0 < \delta < \pi$. A form of greater intuitive appeal is

$$W_{k,m}(z) \sim z^k e^{-z/2}\left[1 + \frac{\{4m^2 - (2k - 1)^2\}}{4.(1!)z}\right.$$
$$\left. + \frac{\{4m^2 - (2k - 1)^2\}\{4m^2 - (2k - 3)^2\}}{4^2.(2!)z^2} + \dots\right]. \quad (7.8.2)$$

From (7.8.2) and (7.7.2) it follows that $U(a; c; z)$ is of order $|z|^{-a}$ as $|z| \to \infty$, with $-\pi < \arg z < \pi$.

We pointed out in section 7.5 that $U(a; c; z)$ is not in general single-valued. Thus although $U(a; c; z)$ is only of algebraic order as $|z| \to \infty$, when $-\pi < \arg z < \pi$, it appears from (7.5.3) that $U(a; c; ze^{2\pi i})$ is exponentially large. This must be true since ${}_1F_1(a; c; z)$ is not proportional to $U(a; c; z)$ and so is exponentially large. It follows that the only solution of the confluent hypergeometric equation which can have algebraic behaviour at infinity regardless of possible circuits of the origin is $U(a; c; z)$, provided

$$\frac{\Gamma(a)\,\Gamma(1-c)(1-\mathrm{e}^{-2ci\pi})}{\Gamma(1+a-c)} \quad \text{or} \quad \frac{2\pi\mathrm{i}\,\Gamma(a)\,\mathrm{e}^{-c\pi\mathrm{i}}}{\Gamma(c)\,\Gamma(1+a-c)}$$

vanishes. This occurs only when c or $1 + a - c$ is a negative integer.

7.9 The second solution

In section 7.7 it was pointed out that $W_{-k,m}(-z)$ gave a second solution of the Whittaker equation. There is a difficulty which must be considered. Equation (7.7.3) defines $W_{k,m}(z)$ in the sector $|\arg z| \leqslant \pi - \delta$, and so if we are to use this integral to express the second solution we must have $|\arg(-z)| < \pi$. This means that if $0 < \arg z \leqslant \pi$ we must take $-z = z\,\mathrm{e}^{-\mathrm{i}\pi}$, and if $-\pi \leqslant \arg z < 0$ we must take $-z = z\,\mathrm{e}^{\mathrm{i}\pi}$. This still leaves z undefined when $\arg z = 0$.

It is probably easier to appreciate the situation if we use equations (7.7.2) and (7.4.5). Thus

$$W_{-k,m}(-z) = \mathrm{e}^{z/2}\left\{\frac{\Gamma(-2m)}{\Gamma(k-m+\frac{1}{2})}(-z)^{m+1/2} \times {}_1F_1(k+m+\tfrac{1}{2}; 1+2m; -z) + \frac{\Gamma(2m)}{\Gamma(k+m+\frac{1}{2})}(-z)^{1/2-m}{}_1F_1(k-m+\tfrac{1}{2}; 1-2m; -z)\right\}. \quad (7.9.1)$$

We have already seen that ${}_1F_1(a; c; z)$ is an entire function of z, so that the many-valued character of $W_{-k,m}(-z)$ arises entirely from the factors $(-z)^{m+1/2}$ and $(-z)^{1/2-m}$. When m is half an odd integer, $W_{-k,m}(z)$ will therefore be single-valued.

Define

$$W_{-k,m}{}^{+}(-x) = \lim_{\theta\to\pi} W_{-k,m}(x\,\mathrm{e}^{\mathrm{i}\theta}) \text{ and } W_{-k,m}{}^{-}(-x) = \lim_{\theta\to\pi} W_{-k,m}(x\,\mathrm{e}^{-\mathrm{i}\theta}),$$

where x is real and positive. From (7.9.1) we see that

$$\frac{1}{2\mathrm{i}}\{W_{-k,m}{}^{+}(-x) - W_{-k,m}{}^{-}(-x)\} = \mathrm{e}^{x/2}\left\{\frac{\Gamma(-2m)\sin(m+\frac{1}{2})\pi}{\Gamma(k-m+\frac{1}{2})}x^{m+1/2}{}_1F_1(k+m+\tfrac{1}{2}; 1+2m; -x) + \frac{\Gamma(2m)\sin(\frac{1}{2}-m)\pi}{\Gamma(k+m+\frac{1}{2})}{}_1F_1(k-m+\tfrac{1}{2}; 1-2m; -x)\right\}. \quad (7.9.2)$$

The right-hand side of (7.9.2) is itself a solution of the Whittaker equation.

We have already observed that the second solution is exponentially large as $|z| \to \infty$ in the right-hand half plane away from the real axis.

It follows that all solutions of equation (7.1.6) behave like e^z as $|z| \to \infty$ in the right-hand half plane, with the exception of $U(a; c; z)$, which we saw in the previous section behaves like z^{-a}.

Equations (7.4.5), (7.7.2) and (7.9.1) express the two independent solutions of the Whittaker equation in terms of the solutions of a confluent hypergeometric equation. We may use these equations to express the confluent hypergeometric function in terms of the Whittaker function

$$\Gamma(1-c)\sin c\pi F(a; c; z) = \pi z^{-c/2}\left\{e^{\pm a\pi i}\frac{W_{k,m}(z)}{\Gamma(c-a)} + e^{\mp k\pi i}\frac{W_{-k,m}(-z)}{\Gamma(a)}\right\}, \quad (7.9.3)$$

where in the last bracket $k = c/2 - a$, $m = (c-1)/2$, and the upper sign in the exponent is to be taken if $0 < \arg z < \pi$, and the lower sign if $-\pi < \arg z < 0$.

EXERCISES

1. Establish the following recurrence relations.

   (a) $c\{{}_1F_1(a+1; c: z) - {}_1F_1(a; c; z)\} = z\,{}_1F_1(a+1; c+1; z)$.

   (b) $c(c-1)\{{}_1F_1(a; c; z) - {}_1F_1(a; c-1; z)\} = -zaF(a+1; c+1; z)$.

   (c) $c\dfrac{d}{dz}{}_1F_1(a; c; z) = a\,{}_1F_1(a+1; c+1; z)$.

2. If $\phi(a; c; z) = {}_1F_1(a+1; c; z)/{}_1F_1(a; c; z)$ and $\psi(a; c; z) = {}_1F_1(a; c+1; z)/{}_1F_1(a; c; z)$, show that

$$(z/c)\psi(a+1; c; z) = 1 - \frac{1}{\phi(a; c; z)}$$

   and

$$a\phi(a; c+1; z) = \frac{c}{\psi(a; c; z)} - (c-a).$$

   Hence or otherwise show that in the usual notation for continued fractions

$$a\phi(a; c; z) = \frac{1}{1-}\,\frac{z}{(c-a-1)+}\,\frac{a+1}{1-}\,\frac{z}{(c-a-1)+}\,\frac{a+2}{1-}\cdots$$

3. Show that by changing both dependent and independent variables the Laplace linear differential equation

$$(a_0z + b_0)\frac{d^2w}{dz^2} + (a_1z + b_1)\frac{dw}{dz} + (a_2z + b_2)w = 0$$

   can be reduced in general to the form for the confluent hypergeometric equation.

4. Show that if $\mathscr{R}(s) > 1$, $\mathscr{R}(\nu) > 0$,

$$s^\nu \int_0^\infty e^{-st}t^{\nu-1}{}_1F_1(a; c; t)dt = \Gamma(\nu)\,{}_2F_1(a, \nu; c; s^{-1}).$$

5. Establish the following relations.

(a) $$\mathrm{e}(\mu, z) \equiv \sum_{n=0}^{\infty} \frac{z^n}{\Gamma(\mu+n+1)} = \frac{1}{\Gamma(\mu+1)} {}_1F_1(1; \mu+1; z),$$

$$\mathscr{R}(\mu+1) > 0.$$

(b) $$E_\nu(z) \equiv \int_1^{\infty} \frac{\mathrm{e}^{-zt}}{t^\nu}\, \mathrm{d}t = \Gamma(1-\nu)z^{\nu-1} + \frac{\mathrm{e}^{-z}}{\nu-1} {}_1F_1(1; 2-\nu; z),$$

$\nu > 0$ and not an integer, $\mathscr{R}(z) > 0$.

6. Show that

$$ {}_1F_1(1; c; z) = (1-c)\, \mathrm{e}^{z/2} z^{-c/2} W_{m-\frac{1}{2},m}(z) + \Gamma(c) z^{1-c}\, \mathrm{e}^z,$$

where $2m + 1 = c$ and hence deduce the following asymptotic series.

(a) $$\mathrm{e}(\mu, z) \sim z^{-\mu}\, \mathrm{e}^z - \frac{1}{\Gamma(\mu) z}\left\{1 + \frac{\mu-1}{z} + \frac{(\mu-1)(\mu-2)}{z^2} + \ldots\right\}.$$

(b) $$E_\nu(z) \sim \frac{\mathrm{e}^{-z}}{z}\left\{1 - \frac{\nu}{z} + \frac{\nu(\nu+1)}{z^2} - \ldots\right\}.$$

Show also that if the incomplete gamma function $\int_0^z \mathrm{e}^{-t} t^{\zeta-1}\, \mathrm{d}t$, $\mathscr{R}(z) > 0$, $\mathscr{R}(\zeta) > 0$, is denoted by $\gamma(\zeta, z)$, then

$$\gamma(\zeta, z) \sim \Gamma(\zeta) - \mathrm{e}^{-z} z^{\zeta-1}\left\{1 + \frac{\zeta-1}{z} + \frac{(\zeta-1)(\zeta-2)}{z^2} + \ldots\right\}.$$

7. The Falkner–Skan differential equation, which occurs in boundary-layer theory, is

$$\frac{\mathrm{d}^3 f}{\mathrm{d}\eta^3} + f(\eta)\frac{\mathrm{d}^2 f}{\mathrm{d}\eta^2} = \beta\left\{\left(\frac{\mathrm{d}f}{\mathrm{d}\eta}\right)^2 - 1\right\}.$$

The solutions which are of practical interest must satisfy the condition that as $\eta \to +\infty$, $\mathrm{d}f/\mathrm{d}\eta - 1$ tends exponentially to zero. Put

$$f(\eta) = a + \eta + g(\eta)$$

and hence, assuming that g and all its derivatives are very small compared with unity, show that such a solution exists.

CHAPTER 8

The Legendre Functions

8.1 The generating function

Consider

$$V(z, h) = (1 - 2zh + h^2)^{-1/2}. \tag{8.1.1}$$

If we restrict z to the interior of some circle $|z| < R$, where the choice of R is arbitrary, and impose the bound $|h| < (1 + R^2)^{1/2} - R$, then

$$|2zh - h^2| \leqslant 2|z||h| + |h|^2 < 1.$$

Thus

$$V(z, h) = 1 + \sum_{k=1}^{\infty} A_k(2zh - h^2)^k \tag{8.1.2}$$

where

$$A_k = \frac{(2k)!}{2^{2k}(k!)^2}. \tag{8.1.3}$$

Under these conditions the series (8.1.2) converges absolutely and we may expand the individual terms $h^k(2z - h)^k$ and rearrange the series in powers of h. The coefficient of h^n is

$$A_n(2z)^n - (n-1)A_{n-1}(2z)^{n-2} \dots$$

The final term is the constant $(-1)^{n/2}A_{n/2}$ if n is even, but $(-1)^{(n-1)/2}(n+1)A_{(n+1)/2}z$ if n is odd. Thus we may write

$$V(z, h) = (1 - 2zh + h^2)^{-1/2} = \sum_{n=0}^{\infty} P_n(z)h^n, \tag{8.1.4}$$

where the $P_n(z)$ are polynomials in z of degree n and are even or odd functions of z according as n is even or odd, respectively. These polynomials are called the Legendre polynomials.

From the process described above we can write down an explicit form for the Legendre polynomials, namely

$$P_n(z) = \sum_{r=0}^{m} (-1)^r \frac{\{(2n-2r)!\}z^{n-2r}}{2^n r!\{(n-r)!\}\{(n-2r)!\}} \tag{8.1.5}$$

where $m = [n/2]$, the integer part of $n/2$.

We may use (8.1.5) to write down the Legendre polynomials of lowest order.

$P_0(z) = 1$; $P_1(z) = z$; $P_2(z) = \frac{1}{2}(3z^2 - 1)$; $P_3(z) = \frac{1}{2}(5z^3 - 3z)$;
$P_4(z) = \frac{1}{8}(35z^4 - 30z^2 + 3)$; $P_5(z) = \frac{1}{8}(63z^5 - 70z^3 + 15z)$.

8.2 Elementary properties of Legendre polynomials

We may deduce some simple properties immediately from the generating function (8.1.4). If we substitute $z = 1$, we see that

$$V(1, h) = (1-h)^{-1} = \sum_{n=0}^{\infty} h^n = \sum_{n=0}^{\infty} P_n(1)h^n$$

and so for every n,

$$P_n(1) = 1. \tag{8.2.1}$$

Similarly we deduce that

$$P_n(-1) = (-1)^n. \tag{8.2.2}$$

We may also use the same idea to find the value of $P_n(0)$. In this case

$$V(0, h) = (1 + h^2)^{-1/2} = \sum_{k=0}^{\infty} (-1)^k A_k h^{2k} = \sum_{n=0}^{\infty} P_n(0)h^n,$$

where A_k is defined by (8.1.2). Thus we deduce that if n is odd, $P_n(0) = 0$. This also follows immediately from our knowledge that $P_n(z)$ is an odd function of z. If, however, n is even,

$$P_n(0) = (-1)^{n/2} A_{n/2}. \tag{8.2.3}$$

We also notice that if $|h| < 1$

$$\begin{aligned} V(\cos\theta, h) &= (1 - h\,e^{i\theta})^{-1/2}(1 - h\,e^{-i\theta})^{-1/2} \\ &= \left(\sum_{k=0}^{\infty} A_k h^k\,e^{ik\theta}\right)\left(\sum_{l=0}^{\infty} A_l h^l\,e^{-il\theta}\right). \\ &= \sum_{n=0}^{\infty}\left(\sum_{r=0}^{n} A_{n-r}A_r\,e^{i(n-r)\theta}\right)h^n. \end{aligned}$$

Thus we deduce that

$$P_n(\cos\theta) = 2A_n \cos n\theta + 2A_{n-1}A_1 \cos(n-2)\theta + \dots \tag{8.2.4}$$

Note that if n is even the final term is $(A_{n/2})^2$, while if n is odd the final term is $2A_{(n+1)/2}A_{(n-1)/2}\cos\theta$.

The important inequality

$$|P_n(\cos\theta)| \leqslant 2A_n + 2A_{n-1}A_1 + \ldots = P_n(1) = 1 \qquad (8.2.5)$$

follows immediately from (8.2.1) and (8.2.4).

8.3 Recurrence relations between Legendre polynomials

From partial differentiation of (8.1.1) we have

$$\frac{\partial V}{\partial h} = (z-h)V^3 \quad \text{and} \quad \frac{\partial V}{\partial z} = hV^3.$$

Thus

$$(h-z)V + (1-2zh+h^2)\frac{\partial V}{\partial h} = 0$$

and

$$h\frac{\partial V}{\partial h} + (h-z)\frac{\partial V}{\partial z} = 0.$$

From (8.1.4) we may find equivalent expansions for the partial derivatives of $V(z, h)$, and if these and (8.1.4) are substituted into the equations above, we derive by equating to zero the coefficients of h^n that

$$(n+1)P_{n+1}(z) - (2n+1)zP_n(z) + nP_{n-1}(z) = 0 \qquad (8.3.1)$$

and

$$nP_n(z) + P_{n-1}'(z) - zP_n'(z) = 0. \qquad (8.3.2)$$

Further

$$\frac{\partial V}{\partial h} = \sum_{n=0}^{\infty} nh^{n-1}P_n(z),$$

so that

$$\sum_{n=0}^{\infty}(2n+1)h^nP_n(z) = \{2h(z-h) + 1 - 2zh + h^2\}V^3$$

$$= (1-h^2)(1-2zh+h^2)^{-3/2}. \qquad (8.3.3)$$

This result has important applications in mathematical physics.

8.4 Other properties of Legendre polynomials

The properties with which we are concerned in this section may all be deduced from an alternative representation of the polynomials, which we establish first.

Since

$$(z^2-1)^n = \sum_{r=0}^{n}(-1)^r\frac{n!z^{2n-2r}}{r!\{(n-r)!\}},$$

it follows that

$$\frac{d^n}{dz^n}(z^2-1)^n = \sum_{r=0}^{[n/2]}(-1)^r \frac{n!\{(2n-2r)!\}z^{n-2r}}{r!\{(n-r)!\}\{(n-2r)!\}}.$$

If we compare this with (8.1.5) we derive Rodrigues's formula

$$2^n n! P_n(z) = \frac{d^n}{dz^n}(z^2-1)^n. \tag{8.4.1}$$

Trivially $(z^2-1)^n$ is a polynomial of degree $2n$ with n real zeros at 1 and n real zeros at -1. By Rolle's theorem, therefore, $(d/dz)(z^2-1)^n$ has $2n-1$ real zeros, $n-1$ of which are at 1, and $n-1$ at -1, and one lies between them and must be at $z = 0$. If we apply the theorem again we find that $(d^2/dz^2)(z^2-1)^n$ has $n-2$ zeros at 1 and $n-2$ zeros at -1, and two further zeros, one in the interval $(-1, 0)$, and the other in $(0, 1)$. We may continue this argument and so deduce that $P_n(z)$ has n real zeros, all lying between -1 and 1. From the recurrence relation (8.3.1) it follows that $P_{n-1}(z)$ and $P_{n+1}(z)$ have opposite signs at a zero of $P_n(z)$. Thus if we note (8.2.1) and (8.2.2) we see that the zeros of $P_{n-1}(z)$ must separate those of $P_n(z)$.

From (8.4.1)

$$\begin{aligned}2^{n+m}(n!)(m!)\int_{-1}^{1} P_n(x)P_m(x)\,dx &= \int_{-1}^{1}\left\{\frac{d^n}{dx^n}(x^2-1)^n\right\}\left\{\frac{d^m}{dx^m}(x^2-1)^m\right\}dx \\ &= \left\{\frac{d^{n-1}}{dx^{n-1}}(x^2-1)^n\right\}\left\{\frac{d^m}{dx^m}(x^2-1)^m\right\}\Bigg|_{-1}^{1} \\ &\quad - \int_{-1}^{1}\left\{\frac{d^{n-1}}{dx^{n-1}}(x^2-1)^n\right\}\left\{\frac{d^{m+1}}{dx^{m+1}}(x^2-1)^m\right\}dx\end{aligned}$$

after an integration by parts. Now $(d^{n-1}/dx^{n-1})(x^2-1)^n$ contains a factor (x^2-1) and so the integrated part vanishes at both limits. We may integrate again, and continue until we have performed n integrations by parts. At this stage the integral is represented by

$$(-1)^n\int_{-1}^{1}(x^2-1)^n\left\{\frac{d^{m+n}}{dx^{m+n}}(x^2-1)^m\right\}dx.$$

There is no loss of generality if we assume that $m \leqslant n$. If $m < n$, $m+n$ exceeds the degree of $(x^2-1)^m$ and so the integrand vanishes identically.

Hence

$$\int_{-1}^{1} P_n(x)P_m(x)\,dx = 0, \qquad (n \neq m) \tag{8.4.2}$$

which is the orthogonality relation for the Legendre polynomials. If $m = n$, then

$$\frac{d^{2n}}{dx^{2n}}(x^2-1)^n = (2n)!$$

and

$$\int_{-1}^{1}(x^2-1)^n\,dx = 2\int_0^1 (x^2-1)^n\,dx = (-1)^n\int_0^1 (1-y)^n y^{-1/2}\,dy$$
$$= B(n+1, \tfrac{1}{2})$$

on putting $y = x^2$. Hence

$$\int_{-1}^{1}\{P_n(x)\}^2\,dx = \frac{\{(2n)!\}n!\,\Gamma(\frac{1}{2})}{2^{2n}(n!)^2\,\Gamma(n+\frac{3}{2})} = \frac{2}{2n+1}, \tag{8.4.3}$$

which is the normalizing relation.

8.5 The expansion theorem for polynomials

Since $P_n(z)$ is a polynomial of degree n, corresponding to any other polynomial $f(z)$ of degree n we can choose a constant a_n such that

$$f(z) - a_nP_n(z)$$

is a polynomial of degree $n - 1$. We can then select the constant a_{n-1} such that

$$f(z) - a_nP_n(z) - a_{n-1}P_{n-1}(z)$$

is a polynomial of degree $n - 2$. Continuing this process we can select successively the constants $a_n, a_{n-1}, \ldots a_1$, such that

$$f(z) - \sum_{k=1}^{n} a_kP_k(z)$$

is a polynomial of degree zero, a constant. This we may designate a_0, and since $P_0(z) = 1$, it may also be written $a_0P_0(z)$. Hence if $f(z)$ is any polynomial of degree n,

$$f(z) = \sum_{k=0}^{n} a_kP_k(z), \tag{8.5.1}$$

and from (8.4.2) and (8.4.3)

$$2a_k = (2k+1)\int_{-1}^{1} f(x)P_k(x)\,dx. \tag{8.5.2}$$

Let us consider the important case when $f(z) = z^n$. Since $P_k(z)$ is an odd or even function of z according as k is odd or even, it follows from (8.5.2) that $a_k = 0$ when $n - k$ is odd. For $n - k = 2r$ we have

$$2^kk!\int_{-1}^{1} x^nP_k(x)\,dx = \int_{-1}^{1} x^n\left\{\frac{d^k}{dx^k}(x^2-1)^k\right\}dx$$
$$= (-1)^kn(n-1)\ldots(2r+1)\int_{-1}^{1} x^{2r}(x^2-1)^k\,dx,$$
$$= \frac{n!}{(2r)!}\int_0^1 y^{r-1/2}(1-y)^k\,dy,$$
$$= \frac{n!}{(2r)!}B(r+\tfrac{1}{2}, k+1).$$

Thus

$$a_k = \frac{2k+1}{2}\int_{-1}^{1} x^n P_k(x)\,\mathrm{d}x = \frac{(2k+1)n!\,\Gamma(r+\frac{1}{2})}{2^{k+1}\{(2r)!\}\,\Gamma(k+r+\frac{1}{2})}$$

$$= \frac{\sqrt{\pi}(2k+1)n!\,\Gamma(k+r+1)}{2^{n+1}(r!)\,\Gamma(k+r+1)\,\Gamma(k+r+\frac{3}{2})}$$

$$= (2k+1)2^k\frac{n!\,\Gamma(k+r+1)}{r!\Gamma(2k+2r+2)}$$

$$= (2k+1)2^k\frac{n![\{\frac{1}{2}(k+n)\}!]}{[\{\frac{1}{2}(n-k)\}!]\{(k+n+1)!\}}. \tag{8.5.3}$$

The combined relations (8.5.1) and (8.5.2) suggest that there may be a larger class of functions than the polynomials which may be represented, at least on the real interval $(-1, 1)$, as a series of Legendre polynomials. We discussed in section 3.10 the completeness of the orthogonal set, and we shall revert to this topic, but from a different standpoint, in section 8.12. We merely note here that if such a class of functions exists, and it must clearly include the polynomials, then a member $f(z)$ of the class has the representation

$$f(z) = \sum_{k=0}^{\infty} a_k P_k(z), \tag{8.5.4}$$

where a_k is still given by (8.5.2).

8.6 The differential equation

If we differentiate the recurrence relation (8.3.1) and subtract n times (8.3.2) from it we obtain

$$P_{n+1}'(z) - (n+1)P_n(z) - zP_n'(z) = 0. \tag{8.6.1}$$

Now replace $n+1$ by n in (8.6.1) and use (8.3.2) to eliminate $P_{n-1}'(z)$. The result is

$$(z^2-1)P_n'(z) - nzP_n(z) - nP_{n-1}(z) = 0. \tag{8.6.2}$$

If we differentiate (8.6.2) and again use (8.3.2) to eliminate $P_{n-1}'(z)$, we derive the differential equation satisfied by $P_n(z)$, namely

$$\frac{\mathrm{d}}{\mathrm{d}z}\left\{(1-z^2)\frac{\mathrm{d}P_n(z)}{\mathrm{d}z}\right\} + n(n+1)P_n(z) = 0. \tag{8.6.3}$$

This leads us to consider the Legendre differential equation

$$\frac{\mathrm{d}}{\mathrm{d}z}\left\{(1-z^2)\frac{\mathrm{d}w}{\mathrm{d}z}\right\} + \nu(\nu+1)w = 0, \tag{8.6.4}$$

where ν is some constant parameter.

This equation has singular points of the first kind at ± 1, and at ∞. Every other point of the complex plane is ordinary. Equations (1.2.5) and (1.2.6) give the series expansions for a pair of independent solutions in the neighbourhood of the origin.

A pair of solutions valid in the neighbourhood of 1 is given by (1.5.5) and (1.5.7), and since there is a natural symmetry between ± 1, a pair valid near -1 has the same structure. We may also find expansions in inverse powers valid for large $|z|$, and these are given by equations (1.7.4) and (1.7.5).

If instead of solutions in series of (8.6.4) we choose to solve the equation by a contour integral of the Euler type, we obtain solutions of the form (2.10.10), provided that the condition (2.10.11) is satisfied.

8.7 The Legendre function of the first kind

We saw in section 2.10 that one suitable choice of contour for an integral solution of (8.6.4) is a simple closed curve enclosing the points $t = z$ and $t = 1$, but excluding $t = -1$. Let us call this contour C_1.

In the particular case when ν is a positive integer n, the integrand is regular at $t = \pm 1$, and has a pole of order $n + 1$ at $t = z$. The integral then takes the form of a Cauchy-type integral for an nth derivative. Hence

$$\frac{n!}{2\pi i}\int_{C_1}\frac{(t^2-1)^n}{(t-z)^{n+1}}\,dt = \frac{d^n}{dz^n}(z^2-1)^n = 2^n n! P_n(z). \tag{8.7.1}$$

We may therefore define the Legendre function of order ν of the first kind by the so-called Schläfli integral

$$P_\nu(z) = \frac{1}{2\pi i}\int_{C_1}\frac{(t^2-1)^\nu\,dt}{2^\nu(t-z)^{\nu+1}}. \tag{8.7.2}$$

This reduces to the Legendre polynomial $P_n(z)$ when ν is a positive integer n. To make the definition by (8.7.2) quite precise we must define the branches of the many-valued functions which occur. We use the fact that C_1 crosses the real axis at a unique point $t_0 > 1$. The branches are prescribed by the conditions $\arg(t_0 - 1) = \arg(t_0 + 1) = 0$, and $|\arg(t_0 - z)| < \pi$.

For all ν it follows from (8.7.2) that

$$P_\nu(1) = \frac{1}{2\pi i}\int_{C_1}\frac{(t+1)^\nu\,dt}{2^\nu(t-1)} = 1. \tag{8.7.3}$$

Hence $P_\nu(z)$ is the uniquely defined solution of (8.6.4) which takes the value unity at $z = 1$. This is just the solution (1.5.5). If we compare this expansion with the expansion of the hypergeometric function (6.2.3), we see that

$$P_\nu(z) = {}_2F_1\{\nu+1, -\nu; 1; \tfrac{1}{2}(1-z)\}. \tag{8.7.4}$$

This result is known as Murphy's formula. It follows from this result, and the properties of the hypergeometric function, that (8.7.4), and therefore by implication (8.7.2), define $P_\nu(z)$ to be a function regular in the complex plane cut from $-\infty$ to -1 along the negative real axis. If we use equation (6.7.4) we see that

$$P_\nu(z) = \frac{\Gamma(2\nu+1)}{\{\Gamma(\nu+1)\}^2}\left(\frac{z-1}{2}\right)^\nu F\{-\nu, -\nu; -2\nu; 2/(1-z)\}$$
$$+ \frac{\tan\nu\pi}{2\pi}\frac{\{\Gamma(\nu+1)\}^2}{\Gamma(2\nu+2)}\left(\frac{z-1}{2}\right)^{-1-\nu} F\{\nu+1, \nu+1; 2\nu+2; 2/(1-z)\}.$$

Hence if $\mathscr{R}(\nu) > 0$, as $|z| \to \infty$

$$P_\nu(z) \sim \frac{\Gamma(2\nu+1)}{2^\nu\{\Gamma(\nu+1)\}^2} z^\nu. \tag{8.7.5}$$

We note incidentally that (8.7.4) is unchanged if ν is replaced by $-1-\nu$, and so

$$P_\nu(z) = P_{-\nu-1}(z). \tag{8.7.6}$$

One of the series (1.2.5) and (1.2.6) terminates if ν is an integer n. In all cases one is an infinite series, and if ν is not an integer both are infinite series. These infinite series have a radius of convergence of unity, and so the functions they represent have a singularity on the circle $|z| = 1$. Now a singularity of a solution of a differential equation must coincide with a singular point of the equation, and the only finite singular points are ± 1. It is clear that both series have singularities at ± 1, as they are both effectively power series in z^2.

The function $P_\nu(z)$ is a linear combination of (1.2.5) and (1.2.6) and we have seen that it is a combination which removes the singularity at 1. On the other hand, the series (1.5.5) has as its circle of convergence $|z-1| = 2$. By the same argument, if ν is not an integer, $P_\nu(z)$ has a singularity at -1. We may further deduce that since the equation is symmetric in $z+1$ and $z-1$, and the second solution at 1 is logarithmic, the singularity of $P_\nu(z)$ at -1 is logarithmic. Thus we see that any solution of (8.6.4) has a logarithmic singularity at either 1 or -1 or both, unless ν is an integer n. In that case one solution is a Legendre polynomial, and so is regular in the whole finite complex plane, but the second solution has singularities at both ± 1.

8.8 Recurrence relations

Let us suppose for the moment that instead of C_1 we take any suitable contour C for the integral. Then it follows that

$$0 = \frac{1}{1+\nu}\int_C \frac{d}{dt}\left\{\frac{(t^2-1)^{\nu+1}}{(t-z)^{\nu+1}}\right\} dt = 2\int_C \frac{t(t^2-1)^\nu\, dt}{(t-z)^{\nu+1}}$$

$$-\int_C \frac{(t^2-1)^{\nu+1}}{(t-z)^{\nu+2}}\,dt - 2\int_C \frac{(t^2-1)^{\nu}}{(t-z)^{\nu}}\,dt$$
$$+ 2z\int_C \frac{(t^2-1)^{\nu}}{(t-z)^{\nu+1}}\,dt - \int_C \frac{(t^2-1)^{\nu+1}}{(t-z)^{\nu+2}}\,dt.$$

In the particular case of the contour C_1 we have from (8.7.2) that

$$\frac{1}{2\pi i}\int_{C_1} \frac{(t^2-1)^{\nu}}{2^{\nu}(t-z)^{\nu}}\,dt = P_{\nu+1}(z) - zP_{\nu}(z). \tag{8.8.1}$$

If we now differentiate this with respect to z we find that

$$P_{\nu+1}'(z) - zP_{\nu}'(z) - (\nu+1)P_{\nu}(z) = 0. \tag{8.8.2}$$

which generalizes the relation (8.6.1).

In a similar way

$$0 = \int_C \frac{d}{dt}\left\{\frac{t(t^2-1)^{\nu}}{(t-z)^{\nu}}\right\}dt = \int_C \frac{(t^2-1)^{\nu}}{(t-z)^{\nu}}\,dt + 2\nu\int_C \frac{t^2(t^2-1)^{\nu-1}}{(t-z)^{\nu}}\,dt$$
$$- \nu\int_C \frac{t(t^2-1)^{\nu}}{(t-z)^{\nu+1}}\,dt$$
$$= (\nu+1)\int_C \frac{(t^2-1)^{\nu}}{(t-z)^{\nu}}\,dt + 2\nu\int_C \frac{(t^2-1)^{\nu-1}}{(t-z)^{\nu}}\,dt$$
$$- \nu z\int_C \frac{(t^2-1)^{\nu}}{(t-z)^{\nu+1}}\,dt.$$

If we take C_1 for the contour and use (8.8.1) we obtain

$$(\nu+1)P_{\nu+1}(z) - (2\nu+1)zP_{\nu}(z) + \nu P_{\nu-1}(z) = 0, \tag{8.8.3}$$

which generalizes (8.3.1).

Now differentiate (8.8.3) and subtract from it $(\nu+1)$ times (8.8.2) and we get

$$\nu P_{\nu}(z) + P_{\nu-1}'(z) - zP_{\nu}'(z) = 0, \tag{8.8.4}$$

which generalizes (8.3.2). Next subtract (8.8.4) from (8.8.3) and we get the symmetric relation

$$P_{\nu+1}'(z) - P_{\nu-1}'(z) - (2\nu+1)P_{\nu}(z) = 0. \tag{8.8.5}$$

We may also obtain the generalization of (8.6.2) if we replace $\nu+1$ by ν in (8.8.2) and add to it z times (8.8.4). This gives

$$(z^2-1)P_{\nu}'(z) + \nu P_{\nu-1}(z) - \nu zP_{\nu}(z) = 0. \tag{8.8.6}$$

Of the five recurrence relations given above, (8.8.3) and any other will be sufficient to deduce the other three. Even these two relations cannot be entirely independent. If we write down the relation (8.8.4) with ν replaced successively by 1, 2, . . . N, we obtain a set of N simultaneous linear differential equations. It is therefore only necessary for the values at a single point of the functions which satisfy these equations to satisfy also the relation (8.8.3) for the solutions to be uniquely determined. They will, of course, be the Legendre polynomials.

We observe finally that the recurrence relations did not depend on the choice of contour for C. Solutions defined by other contours will also satisfy the same recurrence relations, unless constants depending on the parameter are absorbed into their definition.

8.9 The Laplace integrals for $P_\nu(z)$

Suppose that z lies within the right-hand half-plane. Then $|z-1| < |z+1|$, and so the circle $|t-z| = |z^2-1|^{1/2}$ includes $t = 1$ and excludes $t = -1$. It is therefore a possible choice for C_1. On this circle we may put $t = z + (z^2-1)^{1/2}\,\mathrm{e}^{\mathrm{i}\phi}$, with a suitable choice for the branch of $(z^2-1)^{1/2}$. Then

$$\begin{aligned} t^2 - 1 &= z^2 - 1 + 2z(z^2-1)^{1/2}\,\mathrm{e}^{\mathrm{i}\phi} + (z^2-1)\,\mathrm{e}^{2\mathrm{i}\phi} \\ &= 2(z^2-1)^{1/2}\{z + (z^2-1)^{1/2}\cos\phi\}\,\mathrm{e}^{\mathrm{i}\phi}. \end{aligned}$$

When these substitutions are introduced into the Schläfli integral, it reduces to

$$\begin{aligned} \pi P_\nu(z) &= \tfrac{1}{2}\int_{-\pi}^{\pi} \{z + (z^2-1)^{1/2}\cos\phi\}^\nu\,\mathrm{d}\phi \\ &= \int_0^\pi \{z + (z^2-1)^{1/2}\cos\phi\}^\nu\,\mathrm{d}\phi. \end{aligned} \tag{8.9.1}$$

This is the first Laplace integral for $P_\nu(z)$. Once again it is necessary to define precisely the branches of the many-valued functions which occur. If $\phi = \phi_0$ gives that point on C_1 where the contour crosses the real axis to the right of z, we impose the restriction

$$|\arg\{z + (z^2-1)^{1/2}\cos\phi_0\}| < \pi.$$

Using (8.7.6) we obtain the second Laplace integral

$$\pi P_\nu(z) = \int_0^\pi \{z + (z^2-1)^{1/2}\cos\phi\}^{-1-\nu}\,\mathrm{d}\phi. \tag{8.9.2}$$

The function $\{z + (z^2-1)^{1/2}\cos\phi\}^\nu$ has a branch point at $z = \mathrm{i}\cot\phi$ or $-\mathrm{i}\cot\phi$, depending on the choice of the branch of $(z^2-1)^{1/2}$. It follows that the integral (8.9.1) has singularities at all points of the imaginary axis, and the result cannot be continued analytically into the left-hand half plane.

8.10 The Legendre function of the second kind

Another suitable choice of contour for the integral (2.10.10) is a closed curve which passes once round $t = 1$ in one direction and once round $t = -1$ in the other, as long as $t = z$ is not included in either loop.

Hence, provided z does not lie on the real axis between -1 and 1, we can define a second solution of the Legendre equation by

$$Q_\nu(z) = \frac{1}{4\mathrm{i} \sin \nu\pi} \int_{C_2} \frac{(t^2 - 1)^\nu}{(z - t)^{\nu+1}} \, \mathrm{d}t. \qquad (8.10.1)$$

The multiplying constant is chosen for convenience, and its motivation will appear subsequently. The contour C_2 is shown in Fig. 8.1. It is

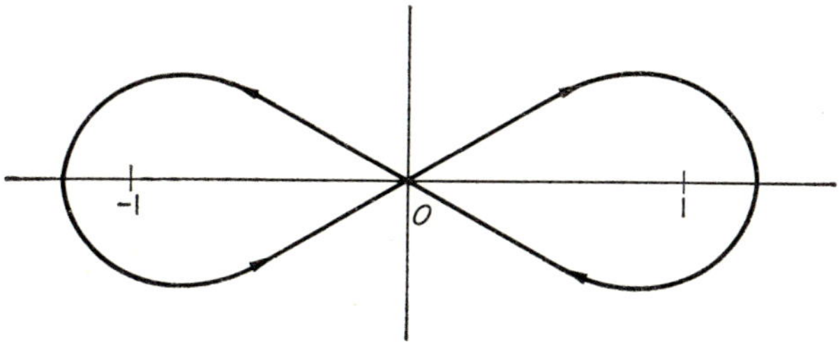

Fig. 8.1

a figure of eight with the loops enclosing $t = \pm 1$, and excluding $t = z$. It will also be shown that to make $Q_\nu(z)$ single-valued the z-plane must be cut from $-\infty$ to 1 along the real axis. In consequence we assume that $|\arg z| < \pi$. The branches of the many-valued functions are to be chosen so that $\arg (t - 1) = \arg (t + 1) = 0$ at the point on C_2 where t is real and $t > 1$. Since we can make C_2 pass through the origin, we also let $\arg (z - t) \to \arg z$ as $t \to 0$.

The contour C_2 can be continuously distorted into the real axis $-1 + \delta \leqslant t \leqslant 1 - \delta$ taken twice and the broken circles $|t \pm 1| = \delta$ (see Fig. 8.2). Let $\nu = \lambda + \mathrm{i}\mu$ and suppose $\lambda + 1 > 0$. The two broken

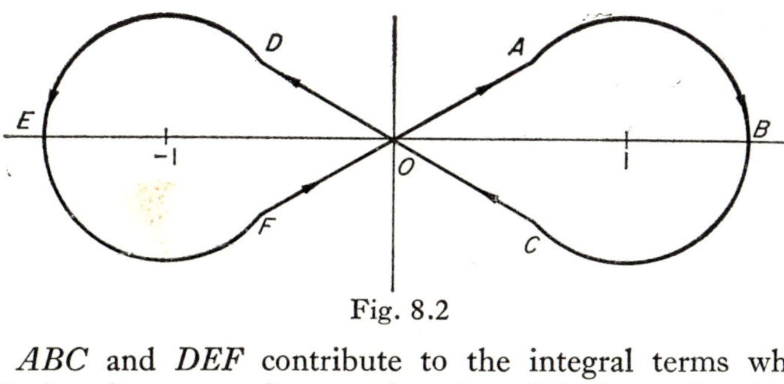

Fig. 8.2

circles ABC and DEF contribute to the integral terms which are $O(\delta^{\lambda+1})$, they do not contribute to the value of the integral as $\delta \to 0$. By assumption, $\arg (t - 1)$ is zero at B, and so $\arg (t - 1)$ is π at A, and is $-\pi$ at C. It follows that on the line FOA

$$t - 1 = |t - 1| \, \mathrm{e}^{\mathrm{i}\pi} = (1 - t) \, \mathrm{e}^{\mathrm{i}\pi}$$

and so

$$(t^2 - 1)^\nu = (1 - t^2)^\nu \, \mathrm{e}^{\mathrm{i}\nu\pi}.$$

By a similar argument, on COD

$$(t^2 - 1)^\nu = (1 - t^2)^\nu \, e^{-i\nu\pi}.$$

Then

$$Q_\nu(z) = \frac{e^{i\nu\pi} - e^{-i\nu\pi}}{4i \sin \nu\pi} \int_{-1}^{1} \frac{(1 - t^2)^\nu}{2^\nu (z - t)^{\nu+1}} \, dt$$

$$= \tfrac{1}{2} \int_{-1}^{1} \frac{(1 - t^2)^\nu}{2^\nu (z - t)^{\nu+1}} \, dt. \tag{8.10.2}$$

We see that $Q_\nu(z)$ satisfies the same recurrence relations as $P_\nu(z)$. If $|z| > 1$, we may write (8.10.2) as

$$Q_\nu(z) = (2z)^{-1-\nu} \int_{-1}^{1} (1 - t^2)^\nu (1 - t/z)^{-1-\nu} \, dt,$$

$$= \frac{(2z)^{-1-\nu}}{\Gamma(1 + \nu)} \sum_{k=0}^{\infty} \frac{\Gamma(\nu + 1 + k)}{k!} \left\{ \int_{-1}^{1} (1 - t^2)^\nu t^k \, dt \right\} z^{-k}.$$

We considered integrals of this form in section 8.5. If k is odd it vanishes, while if $k = 2r$ it reduces to $B(r + \tfrac{1}{2}, \nu + 1)$. Hence

$$Q_\nu(z) = (2z)^{-1-\nu} \sum_{r=0}^{\infty} \frac{\Gamma(r + \tfrac{1}{2}) \, \Gamma(\nu + 1 + 2r)}{(2r)! \, \Gamma(\nu + r + \tfrac{3}{2})} z^{-2r}. \tag{8.10.3}$$

If we apply the duplication formula to expand $(2r)!$ and $\Gamma(\nu + 1 + 2r)$, we may rewrite this in the form

$$Q_\nu(z) = \frac{\pi^{1/2} \, \Gamma(\nu + 1)}{\Gamma(\nu + \tfrac{3}{2})(2z)^{\nu+1}} \, {}_2F_1\{\tfrac{1}{2}(\nu + 1), \tfrac{1}{2}\nu + 1; \nu + \tfrac{3}{2}; z^{-2}\}. \tag{8.10.4}$$

We may alternatively rewrite (8.10.3) in the form

$$Q_\nu(z) = 2^\nu \sum_{r=0}^{\infty} \frac{\Gamma(\nu + r + 1) \, \Gamma(\nu + 2r + 1)}{r! \, \Gamma(2\nu + 2r + 2)} z^{-1-\nu-2r}. \tag{8.10.5}$$

It appears immediately from (8.10.4) that the complex plane must be cut from $-\infty$ to 1 along the real axis, as we stated above. It is also apparent that (8.10.4) defines $Q_\nu(z)$ even when the condition $\mathscr{R}(\nu + 1) > 0$ is not satisfied.

From (8.10.4) it follows that for $|z| \gg 1$

$$Q_\nu(z) \sim \frac{\pi^{1/2} \, \Gamma(\nu + 1)}{\Gamma(\nu + \tfrac{3}{2})(2z)^{\nu+1}}. \tag{8.10.6}$$

The Abel–Liouville theorem [7] yields the Wronskian of $P_\nu(z)$ and $Q_\nu(z)$ in the form

$$P_\nu'(z) Q_\nu(z) - P_\nu(z) Q_\nu'(z) = A(z^2 - 1)^{-1},$$

for some constant A. If, however, we use (8.7.5) and (8.10.6), we derive immediately $A = 1$.

We notice that if ν is an integer, (8.10.3) and (8.10.5) are Laurent series. In this case $Q_n(z)$ is regular in the finite plane cut from -1 to 1. In particular if n is a positive integer, $Q_n(z)$ is regular in the neighbourhood of ∞.

If we substitute

$$t = \{(z+1)^{1/2}\,\mathrm{e}^{\theta} - (z-1)^{1/2}\}/\{(z+1)^{1/2}\,\mathrm{e}^{\theta} + (z-1)^{1/2}\}$$

into (8.10.2), we derive the Laplace integral for $Q_\nu(z)$, which is

$$Q_\nu(z) = \tfrac{1}{2}\int_0^\infty \{z + (z^2-1)^{1/2}\cosh\theta\}^{-1-\nu}\,\mathrm{d}\theta, \qquad (8.10.7)$$

where it is assumed that $\arg\{z+(z^2-1)^{1/2}\}$ takes its principal value.

8.11 Neumann's integral

Let n be a positive integer and consider

$$\int_{-1}^{1}\frac{P_n(t)\,\mathrm{d}t}{z-t}.$$

This defines a function of z which is regular in the complex plane cut from -1 to 1, and is regular at ∞.

Suppose now $|z|>1$, so that, since $|t|\leqslant 1$,

$$\int_{-1}^{1}\frac{P_n(t)\,\mathrm{d}t}{z-t} = \sum_{k=0}^{\infty}\left\{\int_{-1}^{1} t^k P_n(t)\,\mathrm{d}t\right\}z^{-1-k}.$$

Integrals of the form that appear in the bracket were evaluated in section 8.5, though with k and n reversed. Thus the series on the right has the form

$$2^{n+1}\sum_{r=0}^{\infty}\frac{\{(n+2r)!\}\{(n+r)!\}}{r!\{(2n+2r+1)!\}}z^{-n-1-2r}.$$

If we compare this with (8.10.5) it follows that if $|z|>1$

$$Q_n(z) = \tfrac{1}{2}\int_{-1}^{1}\frac{P_n(t)\,\mathrm{d}t}{z-t}. \qquad (8.11.1)$$

Both sides of this equation represent functions regular in the cut plane, and regular at infinity. We may therefore continue this equation to the whole cut plane.

The equation (8.11.1) gives a convenient method for obtaining explicit expressions for some of the $Q_n(z)$. For example,

$$Q_0(z) = \tfrac{1}{2}\int_{-1}^{1}\frac{\mathrm{d}t}{z-t} = \tfrac{1}{2}\log\left(\frac{z+1}{z-1}\right)$$

and

$$Q_1(z) = \tfrac{1}{2}\int_{-1}^{1}\frac{t\mathrm{d}t}{z-t} = \frac{z}{2}\int_{-1}^{1}\frac{\mathrm{d}t}{z-t} - 1 = zQ_0(z) - 1.$$

We may always write

$$Q_n(z) = \frac{P_n(z)}{2}\int_{-1}^{1}\frac{\mathrm{d}t}{z-t} + \tfrac{1}{2}\int_{-1}^{1}\frac{P_n(t) - P_n(z)}{z-t}\,\mathrm{d}t,$$

and the latter term is a polynomial in z of degree $n - 1$. Hence

$$Q_n(z) = P_n(z)Q_0(z) + p_n(z), \tag{8.11.2}$$

where $p_n(z)$ is a polynomial in z of degree $n - 1$. If n is even, $p_n(z)$ is an odd function of z, and vice versa.

8.12 A general expansion theorem

In section 8.5 we discussed the expansion of a polynomial as a finite series of Legendre polynomials. We also discussed a more general theorem in section 3.10. In this section we use the result of the previous section to obtain a result of wider application than that of section 8.5.

Suppose that $|\zeta| > |z|$ and also $|\zeta| > 1$. We have the algebraic identity

$$\zeta^{-1}\sum_{k=0}^{n-1}\left(\frac{z}{\zeta}\right)^k = (\zeta - z)^{-1}\{1 - (z/\zeta)^n\}.$$

The left-hand side is a polynomial in z, and so by our earlier result it can be represented in the form

$$\sum_{k=0}^{n-1} F_k(\zeta)P_k(z).$$

It also follows from the earlier analysis that

$$F_k(\zeta) = \frac{2k+1}{2}\left\{\int_{-1}^{1}\frac{P_k(t)\,\mathrm{d}t}{\zeta - t} - \zeta^{-n}\int_{-1}^{1}\frac{t^nP_k(t)\,\mathrm{d}t}{\zeta - t}\right\}.$$

By our hypothesis both $|z/\zeta| < 1$ and $|1/\zeta| < 1$, and so as $n \to \infty$, $|z/\zeta|^n \to 0$ and

$$\left|\zeta^{-n}\int_{-1}^{1}\frac{t^nP_k(t)\,\mathrm{d}t}{\zeta - t}\right| \leqslant 2|\zeta|^{-n}\{|\zeta| - 1\}^{-1} \to 0.$$

Thus in the limit

$$\frac{1}{\zeta - z} = \sum_{k=0}^{\infty}\frac{2k+1}{2}\left\{\int_{-1}^{1}\frac{P_k(t)\,\mathrm{d}t}{\zeta - t}\right\}P_k(z),$$

$$= \sum_{k=0}^{\infty}(2k+1)Q_k(\zeta)P_k(z), \tag{8.12.1}$$

by (8.11.1). This result is due to Heine.

Now let $f(z)$ be any function regular in a domain which has in its interior the circle $|z| \leqslant \rho$, where ρ is any constant greater than unity. We call this circle C. By the Cauchy integral formula, if $|z_0| < \rho$, then

$$f(z_0) = \frac{1}{2\pi i}\int_C \frac{f(\zeta)\,d\zeta}{\zeta - z_0}$$

$$= \sum_{k=0}^{\infty} a_k P_k(z_0), \tag{8.12.2}$$

where from (8.12.1)

$$a_k = \frac{2k+1}{2\pi i}\int_C f(\zeta)Q_k(\zeta)\,d\zeta.$$

By definition $f(\zeta)$ is regular in and on C, while $Q_k(\zeta)$ is regular away from the cut from -1 to 1. If we define

$$Q_k^{+}(t) = \lim_{\varepsilon\to+0} Q_k(t + i\varepsilon) \quad\text{and}\quad Q_k^{-}(t) = \lim_{\varepsilon\to+0} Q_k(t - i\varepsilon)$$

for $-1 < t < 1$, then

$$a_k = -\frac{2k+1}{2\pi i}\int_{-1} f(t)\{Q_k^{+}(t) - Q_k^{-}(t)\}\,dt.$$

In fact we observe from (8.11.2) that $Q_k(t)$ is only many-valued because of the $Q_0(t)$ which it contains implicitly. Hence

$$Q_k^{+}(t) - Q_k^{-}(t) = P_k(t)\{Q_0^{+}(t) - Q_0^{-}(t)\} = -i\pi P_k(t).$$

Finally,

$$a_k = \frac{2k+1}{2}\int_{-1}^{1} f(t)P_k(t)\,dt. \tag{8.12.3}$$

8.13 The associated Legendre functions

If x is real and in the interval $(-1, 1)$, and m is a positive integer, Ferrer has defined an associated Legendre function by the relation

$$P_\nu^m(x) = (1 - x^2)^{m/2}\frac{d^m P_\nu(x)}{dx^m}. \tag{8.13.1}$$

Since $P_\nu(x)$ satisfies the differential equation (8.6.4), it follows from the definition (8.13.1) that $P_\nu^m(x)$ satisfies the associated Legendre equation

$$\frac{d}{dx}\left\{(1 - x^2)\frac{dy}{dx}\right\} + \{\nu(\nu+1) - m^2(1 - x^2)^{-1}\}y = 0. \tag{8.13.2}$$

It is seen from (8.13.1) that if ν is an integer n, then $P_n^m(x) = 0$ if

$m > n$. It is also seen that $P_\nu^0(z) = P_\nu(z)$. The functions of simplest form are

$$P_1^1(x) = (1 - x^2)^{1/2},$$
$$P_2^1(x) = 3(1 - x^2)^{1/2}x, \qquad P_2^2(x) = 3(1 - x^2),$$
$$P_3^1(x) = \tfrac{3}{2}(1 - x^2)^{1/2}(5x^2 - 1), \quad P_3^2(x) = 15(1 - x^2)x,$$
$$P_3^3(x) = 15(1 - x^2)^{3/2}.$$

The arguments that were used to prove that the only solutions of equation (8.6.4) which were bounded at both ± 1 were the Legendre polynomials, which required that ν was an integer n, can be extended to equation (8.13.2). Again we find that a solution can only be bounded at both ± 1 if $\nu = n$, an integer, and this solution is the function $P_n^m(x)$.

Equation (8.13.2) is a Sturm–Liouville equation, and so if m, n and p are all integers and $n \neq p$,

$$\int_{-1}^{1} P_n^m(x) P_p^m(x)\, \mathrm{d}x = 0. \tag{8.13.3}$$

On the other hand

$$2^{2n}(n!)^2 \int_{-1}^{1} \{P_n^m(x)\}^2\, \mathrm{d}x = \int_{-1}^{1} (1 - x^2)^m \left\{\frac{\mathrm{d}^{n+m}}{\mathrm{d}x^{n+m}}(x^2 - 1)^n\right\}^2 \mathrm{d}x$$
$$= (-1)^n \int_{-1}^{1} (x^2 - 1)^n \frac{\mathrm{d}^{n+m}}{\mathrm{d}x^{n+m}}\left\{(x^2 - 1)^m \times \frac{\mathrm{d}^{n+m}}{\mathrm{d}x^{n+m}}(x^2 - 1)^n\right\} \mathrm{d}x$$
$$= \frac{\{(2n)!\}\{(n + m)!\}}{\{(n - m)!\}} \int_{-1}^{1} (1 - x^2)^n\, \mathrm{d}x.$$

Hence

$$\int_{-1}^{1} \{P_n^m(x)\}^2\, \mathrm{d}x = \frac{2\{(n + m)!\}}{(2n + 1)\{(n - m)!\}}. \tag{8.13.4}$$

For values of z not on the real line segment $[-1, 1]$, it is easier to define the branch of the many-valued function $(z^2 - 1)^{m/2}$ rather than that of the term in (8.13.1). It is therefore more convenient to take Hobson's definition of the associated function, which also allows us to define an associated function of the second kind. These are

$$P_\nu^m(z) = (z^2 - 1)^{m/2} \frac{\mathrm{d}^m}{\mathrm{d}z^m} P_\nu(z), \tag{8.13.5}$$

and

$$Q_\nu^m(z) = (z^2 - 1)^{m/2} \frac{\mathrm{d}^m}{\mathrm{d}z^m} Q_\nu(z), \tag{8.13.6}$$

where m is an integer and we take the principal branch of $(z^2 - 1)^{1/2}$.

Then we may derive for $P_\nu^m(z)$ and $Q_\nu^m(z)$ results which generalize some previously derived for $P_\nu(z)$ and $Q_\nu(z)$. We may, for example, obtain directly from (8.7.2) the Schläfli integral

$$P_\nu^m(z) = \frac{\Gamma(\nu+m+1)}{\Gamma(\nu+1)}(z^2-1)^{m/2}\frac{1}{2\pi i}\int_{C_1}\frac{(t^2-1)^\nu\,dt}{2^\nu(t-z)^{\nu+m+1}}. \quad (8.13.7)$$

From (8.13.7) we may obtain the two Laplace integrals, namely

$$P_\nu^m(z) = \frac{\Gamma(\nu+m+1)}{\pi\,\Gamma(\nu+1)}\int_0^\pi \{z + (z^2-1)^{1/2}\cos\phi\}^\nu \cos m\phi\, d\phi, \quad (8.13.8)$$

and

$$P_\nu^m(z) = (-1)^m\frac{\Gamma(\nu+1)}{\pi\,\Gamma(\nu-m+1)}\int_0^\pi \{z + (z^2-1)^{1/2}\cos\phi\}^{-1-\nu} \times \cos m\phi\, d\phi. \quad (8.13.9)$$

We may also use the Schläfli integral to find an equivalent of Murphy's formula. This has the form

$$P_\nu^m(z) = \frac{1}{\Gamma(1-m)}\left(\frac{z+1}{z-1}\right)^{m/2} F\{\nu+1, -\nu; 1-m; \tfrac{1}{2}(1-z)\}. \quad (8.13.10)$$

This result gives a convenient form by which the definition of $P_\nu^m(z)$ can be extended analytically for all m.

8.14 The addition theorem

It follows from Appendix 1 and sections 8.7 and 8.13 that the only solutions of the Laplace equation in spherical polar coordinates (r, θ, ϕ) which are analytic in the whole neighbourhood of $r = 0$, have the form

$$r^n P_n^m(\cos\theta)\cos m\phi \quad \text{or} \quad r^n P_n^m(\cos\theta)\sin m\phi,$$

where n and m are integers.

Let two lines through the origin have the Eulerian angles (θ, ϕ) and (θ', ϕ'), and let the angle between them be Θ. Then

$$\cos\Theta = \cos\theta\cos\theta' + \sin\theta\sin\theta'\cos(\phi-\phi'). \quad (8.14.1)$$

Since $P_n(\cos\Theta)$ is a polynomial in $\cos\Theta$ of degree n, it follows that $P_n(\cos\Theta)$ is a trigonometrical polynomial in $(\phi - \phi')$ of degree n. In consequence it can be represented by a relation

$$P_n(\cos\Theta) = \sum_{m=0}^{n} f_m(\theta, \theta')\cos\{m(\phi-\phi')\},$$

where $f_m(\theta, \theta')$ must be symmetric in θ and θ'. If we multiply through by r^n, the left-hand side is a solution of the Laplace equation, and so also therefore is the right-hand side. Thus this relation must have the form

$$P_n(\cos\Theta) = P_n(\cos\theta)P_n(\cos\theta') + \sum_{m=1}^{n} a_{n,m} P_n{}^m(\cos\theta)P_n{}^m(\cos\theta')\cos m(\phi-\phi'), \quad (8.14.2)$$

where $a_{n,m}$ are constants.

The coefficients $a_{n,m}$ may be obtained by equating the terms on both sides which are of the nth degree in $\cos\theta$, $\sin\theta$, $\cos\theta'$, $\sin\theta'$. Those on the left-hand side are

$$\frac{(2n)!}{2^n\{n!\}^2}\{\cos\theta\cos\theta' + \sin\theta\sin\theta'\cos\psi\}^n,$$

where for convenience we put $\psi = \phi - \phi'$. In this expression we may replace $\cos\psi$ by $\frac{1}{2}(e^{i\psi} + e^{-i\psi})$ and use the multinomial theorem. This gives an expansion

$$\frac{(2n)!}{2^n(n!)}\left[\sum_{j=0}^{[n/2]}\frac{(\cos\theta\cos\theta')^{n-2j}(\sin\theta\sin\theta')^{2j}}{2^{2j}\{(n-2j)!\}(j!)^2} + 2\sum_{m=1}^{n}\left\{\sum_{j=0}^{[(n-m)/2]}\frac{(\cos\theta\cos\theta')^{n-m-2j}(\sin\theta\sin\theta')^{m+2j}}{2^{2j+m}\{(n-m-2j)!\}(j!)\{(j+m)!\}}\right\}\cos m\psi\right]. \quad (8.14.3)$$

This can be slightly simplified. Since we are retaining only the terms of nth degree, we may put

$$(\sin\theta\sin\theta')^{2j} = (1-\cos^2\theta)^j(1-\cos^2\theta')^j = \cos^{2j}\theta\cos^{2j}\theta' + \text{lower powers},$$

and these lower powers may be ignored. The coefficients of $\cos m\psi$ still look unwieldy, but they are the same coefficients which occur in the identity

$$(1+\cos\psi)^n = 1 + 2\sum_{m=1}^{n} b_{n,m}\cos m\psi$$

and can be separately determined. Since the right-hand side in this identity must be the Fourier cosine expansion of the left-hand side,

$$b_{m,n} = \frac{1}{2\pi}\int_{-\pi}^{\pi}(1+\cos\psi)^n\cos m\psi\,d\psi.$$

The substitution $e^{i\psi} = \zeta$ gives

$$b_{m,n} = \mathscr{R}\left\{\frac{1}{2\pi i}\int_{|\zeta|=1}\frac{\zeta^m(\zeta+1)^{2n}\,d\zeta}{(2\zeta)^n}\right\} = \frac{(2n)!}{2^n\{(n+m)!\}\{(n-m)!\}}.$$

Thus (8.14.3) reduces to

$$\frac{(2n)!}{2^n(n!)^2}\left[\frac{(2n)!}{2^n(n!)^2}\cos^n\theta\cos^n\theta' + 2\sum_{m=1}^{n}\frac{(2n)!}{2^n\{(n+m)!\}\{(n-m)!\}}\right.$$
$$\left.\times \sin^m\theta\cos^{n-m}\theta\sin^m\theta'\cos^{n-m}\theta'\cos m\psi\right].$$

Comparing this with the leading terms on the right-hand side of (8.14.2), we can identify the coefficients $a_{n,m}$. The addition formula then takes the form

$$P_n(\cos\Theta) = P_n(\cos\theta)P_n(\cos\theta')$$
$$+ 2\sum_{m=1}^{n}\frac{(n-m)!}{(n+m)!}P_n{}^m(\cos\theta)P_n{}^m(\cos\theta')\cos m(\phi-\phi'), \quad (8.14.4)$$

where Θ is defined by (8.14.1), and the Ferrer definition of $P_n{}^m(x)$ has been used.

Exercises

1. From the identity

$$(1-h+h^2)(1+h+h^2) = 1+h^2+h^4$$

deduce that

$$P_n(\tfrac{1}{2}) = \sum_{r=0}^{2n}(-1)^{n-r}P_{2n-r}(\tfrac{1}{2})P_r(\tfrac{1}{2}).$$

2. Use the substitution $x = \cos\theta$ to show that

$$\int_{-1}^{1}P_n(x)\frac{dx}{(1-x^2)^{1/2}} = \begin{cases}0, & n \text{ odd}\\ 2(A_{n/2})^2, & n \text{ even}\end{cases}$$

where A_n is defined in section 8.1.

3. If $\phi_n(z) = P_n(z)/P_{n-1}(z)$ for $n = 1, 2, 3, \ldots$, show that

$$(n+1)\phi_{n+1}(z) = (2n+1)z - n/\phi_n(z)$$

and hence find a finite continued fraction for $\phi_n(z)$.

4. Show that the nth-order determinant, whose kth row has zeros everywhere but in the $(k-1)$th, kth and $(k+1)$th postions, and wherein the $(k-1)$th element is $(k-1)/k$, the kth is $\{(2k-1)/k\}z$ and the $(k-1)$th element is 1, when expanded is equal to $P_n(z)$. [The same formulae hold for the 1st and 2nd elements of the first row, and the $(n-1)$th and nth elements of the nth row.]

5. Show that each of the following has an expansion of the form $\sum_{k=0}^{n} a_kP_k(z)$;

(a) $P_{n+1}'(z)$; (b) $(1-z^2)P_n''(z)$; (c) $(z-1)^n$.

In each case find an explicit formula for the coefficients.

6. The $(n+1)$th-order determinant $D_m(x)$ has as the jth element of its kth row $1/(2k+2j-1)$ for $j = 1, 2, \ldots, n+1$ and $k = 1, 2, \ldots n$. The jth element of the $(n+1)$th row is $x^{2(j-1)}$. Show that $D_n(x) \propto P_{2n}(x)$. Hence or otherwise find the value of the $(n+1)$th-order determinant with elements $1/(2k+2j-1)$ as the ith element of the kth row for $k = 1, 2, \ldots (n+1)$.
7. Show that

$$\int_0^1 P_n(x)\,dx = (-1)^{(n-1)/2}\frac{A_{(n-1)/2}}{n+1}$$

if n is an odd integer (in the notation of section 8.1) and is zero if n is an even integer.
8. Show that if n and m are integers and $n \geqslant m$, then

$$P_n(z)P_m(z) = \sum_{k=0}^{m} b_{mk}P_{m+n-2k}(z).$$

Verify that $b_{10} = (n+1)/(2n+1)$ and $b_{11} = n/(2n+1)$ and hence verify that

$$b_{mk} = \frac{(2n+2m-4k+1)A_{m-k}A_kA_{n-k}}{(2n+2m-2k+1)A_{n+m-k}}$$

where A_n has the usual meaning.
9. Show that we may represent $e^{\alpha z}$ by a series of Legendre polynomials $\sum_{k=0}^{\infty} a_kP_k(z)$ and that in this series

$$a_k = \frac{\sqrt{\pi}\alpha^k e^{-\alpha}}{2^k\,\Gamma(k+\frac{1}{2})}{}_1F_1(k+1; 2k+2; 2\alpha).$$

10. If we define $P_\nu(z), \mathscr{R}(\nu+1) > 0$ for $|z| < 1$ by the Schläfli integral, and take as the contour C_1 the unit circle indented to the right of both $t = 1$ and $t = -1$, show that

$$P_\nu(0) = \frac{\cos(\nu\pi/2)\,\Gamma\{(\nu+1)/2\}}{\sqrt{\pi}\,\Gamma\{(\nu+2)/2\}}$$

and

$$P_\nu'(0) = 2\frac{\sin(\nu\pi/2)\,\Gamma\{(\nu+2)/2\}}{\sqrt{\pi}\,\Gamma\{(\nu+1)/2\}}.$$

Hence or otherwise deduce that in $|z| < 1$

$$P_\nu(z) = \frac{1}{\sqrt{\pi}}\left[\frac{\Gamma\{(\nu+1)/2\}}{\Gamma\{(\nu+2)/2\}}\cos\frac{\nu\pi}{2}w_0(z) + 2\frac{\Gamma\{(\nu+2)/2\}}{\Gamma\{(\nu+1)/2\}}w_1(z)\right],$$

where $w_1(z)$ and $w_2(z)$ are the series defined by equations (1.2.5) and (1.2.6) respectively.
11. Show that the solution of the Darboux partial differential equation

$$(x+y)\frac{\partial^2\phi}{\partial x\partial y} + \nu\left\{\frac{\partial\phi}{\partial x} + \frac{\partial\phi}{\partial y}\right\} = 0$$

which equals $\{(x_0+y)/(x_0+y_0)\}^{-\nu}$ on $x=x_0$ and $\{(x+y_0)/(x_0+y_0)\}^{-\nu}$ on $y=y_0$ is

$$\phi(x,y)=\left(\frac{x+y}{x_0+y_0}\right)^{-\nu}P_{\nu-1}(1-2\eta)$$

where

$$\eta=\frac{(x-x_0)(y-y_0)}{(x+y)(x_0+y_0)}.$$

12. Show that for integer n

$$Q_n(z)=P_n(z)\int_z^\infty\frac{dt}{(t^2-1)\{P_n(t)\}^2}.$$

13. Defining $p_n(z)$ for integer n by

$$Q_n(z)=P_n(z)Q_0(z)+p_n(z),$$

show that

$$p_n(z)=\sum_{s=0}^{m}\frac{2n-4s-1}{(2s+1)(n-s)}P_{n-2s-1}(z),$$

where $2m=n-1$ if n is odd and $2m=n-2$ if m is even.

14. Show that

$$n\{P_n(z)Q_{n-1}(z)-P_{n-1}(z)Q_n(z)\}=1.$$

Hence or otherwise deduce that if $p_n(z)$ is defined as in exercise 13,

$$p_n(z)+P_n(z)\sum_{r=1}^{n}\{rP_r(z)P_{r-1}(z)\}^{-1}=0.$$

15. Show that for integer m and n with $m\leqslant n$

$$P_m(z)Q_n(z)=\tfrac{1}{2}\int_{-1}^{1}\frac{P_m(t)P_n(t)\,dt}{z-t}.$$

16. Show that, assuming Ferrer's definition,

(a) $$\frac{d}{dz}P_\nu^{\,m}(z)=(1-z^2)^{-1/2}P_\nu^{\,m+1}(z)-\frac{mz}{1-z^2}P_\nu^{\,m}(z),$$

(b) $$(2\nu+1)(1-z^2)^{1/2}P_\nu^{\,m}(z)=P_{\nu+1}^{\,m+1}(z)-P_{\nu-1}^{\,m+1}(z).$$

17. If the Gegenbauer polynomials $C_n^{\,\nu}(z)$ are defined by the generating function

$$(1-2zh+h^2)^{-\nu}=\sum_{n=0}^{\infty}h^nC_n^{\,\nu}(z),$$

find recurrence relations for $C_n^{\,\nu}(z)$ and hence show that $C_n^{\,\nu}(z)$ is a solution of the differential equation

$$(1-z^2)w''-(2\nu+1)zw'+n(n+2\nu)w=0.$$

18. A function satisfies

$$\frac{\partial^2\phi}{\partial x^2}+\frac{\partial^2\phi}{\partial y^2}+\frac{\partial^2\phi}{\partial z^2}=0$$

in the region $R^2 = x^2 + y^2 + z^2 > a^2$, and as $R \to \infty$

$$\phi \sim ARP_1(\cos\theta),$$

where $x^2 + y^2 = R^2 \sin^2\theta$, $z = R\cos\theta$.

On the sphere $R = a$, $\partial\phi/\partial R = 0$. Show that

$$\phi = A\left(R + \frac{a^2}{R^2}\right)P_1(\cos\theta).$$

19. A function satisfies

$$\frac{\partial^2\phi}{\partial x^2}+\frac{\partial^2\phi}{\partial y^2}+\frac{\partial^2\phi}{\partial z^2}=0$$

in the interior of the "cone" $0 < R < \infty$, $0 \leqslant \theta < \alpha < \pi$. As $R \to \infty$, $R\phi \to 0$ and on $\theta = \alpha$, $\phi = e^{-R}$. Show that if $0 < \mathscr{R}(s) < 1$, and

$$\bar{\phi}(s,\theta) = \int_9^\infty R^{-s}\phi(R,\theta)\,dR,$$

then

$$\bar{\phi}(s,\theta) = \frac{\Gamma(1-s)P_{s-1}(\cos\theta)}{P_{s-1}(\cos\alpha)}.$$

The notation is as in exercise 18.

Answers

3. $n\phi_n(z) = (2n-1)z - \dfrac{(n-1)^2}{(2n-3)z-}\;\dfrac{(n-2)^2}{(2n-5)z-}\;\dfrac{(n-3)^2}{(2n-7)z-}\cdots.$

5. (a) $a_k = (2k+1)$ if $(n+k)$ is even and is zero if $(n+k)$ is odd.

   (b) $a_k = 0$ if $(n+k)$ is odd,

   $= 2(2k+1)$, $k = 2, 4, \ldots (n-2)$,

   $= -n(n-1)$, $k = n$.

   (c) $a_k = (-1)^{k+n}2^n \dfrac{(2k+1)(n!)^2}{\{(n+k+1)!\}\{(n-k)!\}}.$

17. $(n+1)C_{n+1}^{\nu}(z) - (2n+2\nu)zC_n^{\nu}(z) + (n-1+2\nu)C_{n-1}^{\nu}(z) = 0,$

$$zC_n^{\nu\prime}(z) - nC_n^{\nu}(z) - C_{n-1}^{\nu\prime}(z) = 0.$$

CHAPTER 9

Bessel Functions

9.1 The generating function

For every finite z the function

$$\exp\left\{\frac{z}{2}\left(t - \frac{1}{t}\right)\right\}$$

represents a function of t which is regular in the whole finite t-plane from which 0 is omitted. It can therefore be expanded in a Laurent series

$$\exp\left\{\frac{z}{2}\left(t - \frac{1}{t}\right)\right\} = \sum_{n=-\infty}^{\infty} t^n J_n(z), \tag{9.1.1}$$

which is absolutely convergent for $0 < |t| < \infty$.

This defines a set of functions of z, depending on an integral parameter n, and these are called the Bessel functions of the first kind, of integral order. Since the left-hand side of (9.1.1) is an entire function of z, for every fixed t within $0 < |t| < \infty$, the functions $J_n(z)$ are entire functions of z.

If in (9.1.1) we replace t by $-1/\tau$, we see that

$$\exp\left\{\frac{z}{2}\left(\tau - \frac{1}{\tau}\right)\right\} = \sum_{m=-\infty}^{\infty} (-1)^n \tau^{-n} J_n(z) = \sum_{m=-\infty}^{\infty} \tau^n J_n(z),$$

and so

$$J_n(z) = (-1)^n J_{-n}(z). \tag{9.1.2}$$

We may use (9.1.1) to obtain explicit expressions for $J_n(z)$. For any finite z and $0 < |t| < \infty$ both the series

$$\exp\left(\tfrac{1}{2}tz\right) = \sum_{n=0}^{\infty} \frac{1}{n!}\left(\frac{tz}{2}\right)^n$$

and

$$\exp\left(-\frac{z}{2t}\right) = \sum_{n=0}^{\infty} \frac{(-1)^n}{n!}\left(\frac{z}{2t}\right)^n$$

converge absolutely and so can be multiplied term by term. The result of this process is

$$J_n(z) = \left(\frac{z}{2}\right)^n \sum_{r=0}^{\infty} \frac{(-1)^r}{\{(n+r)!\}\, r!}\left(\frac{z}{2}\right)^{2r}. \tag{9.1.3}$$

A second expression

$$J_n(z) = \frac{1}{2\pi i}\int_C t^{-n-1} \exp\left\{\frac{z}{2}\left(t - \frac{1}{t}\right)\right\} dt, \tag{9.1.4}$$

where C is any simple Jordan curve with $t = 0$ in its interior, can be obtained if we use the integral formula for a Laurent coefficient.

In particular, take C to be the unit circle on which $t = e^{i\theta}$, and so

$$J_n(z) = \frac{1}{2\pi}\int_{-\pi}^{\pi} \exp(iz\sin\theta - in\theta)\, d\theta.$$

The imaginary part of the integrand is an odd function of θ and so makes no contribution to the integral. The real part is even, and hence

$$J_n(z) = \frac{1}{\pi}\int_0^{\pi} \cos(z\sin\theta - n\theta)\, d\theta. \tag{9.1.5}$$

9.2 Recurrence relations

We may use (9.1.1) to obtain recurrence relations between contiguous Bessel functions. Differentiating both sides of (9.1.1) with respect to t, we obtain

$$\frac{z}{2}\left(1 + \frac{1}{t^2}\right) \exp\left\{\frac{z}{2}\left(t - \frac{1}{t}\right)\right\} = \sum_{n=-\infty}^{\infty} n t^{n-1} J_n(z).$$

If we replace the exponential on the left-hand side by its expansion (9.1.1), and then equate coefficients of t^{n-1}, we obtain

$$z\{J_{n-1}(z) + J_{n+1}(z)\} = 2nJ_n(z). \tag{9.2.1}$$

Similarly, if we differentiate (9.1.1) with respect to z, we have

$$\tfrac{1}{2}\left(t - \frac{1}{t}\right) \exp\left\{\frac{z}{2}\left(t - \frac{1}{t}\right)\right\} = \sum_{n=-\infty}^{\infty} t^n J_n'(z)$$

and hence

$$J_{n-1}(z) - J_{n+1}(z) = 2J_n'(z). \tag{9.2.2}$$

From (9.2.1) and (9.2.2) we obtain, by eliminating $J_{n+1}(z)$

$$zJ_{n-1}(z) = nJ_n(z) + zJ_n'(z),$$

that is,

$$z^n J_{n-1}(z) = \frac{d}{dz}\{z^n J_n(z)\}. \tag{9.2.3}$$

Similarly, if we eliminate $J_{n-1}(z)$ we derive

$$z^{-n}J_{n+1}(z) = -\frac{d}{dz}\{z^{-n}J_n(z)\}. \tag{9.2.4}$$

From (9.2.3) and (9.2.4) we may deduce that

$$z^{n+1}J_n(z) = \frac{d}{dz}\{z^{n+1}J_{n+1}(z)\} = -\frac{d}{dz}\left[z^{2n+1}\frac{d}{dz}\{z^{-n}J_n(z)\}\right].$$

On simplification we obtain

$$z^2J_n''(z) + zJ_n'(z) + (z^2 - n^2)J_n(z) = 0. \tag{9.2.5}$$

9.3 The Bessel differential equation

We see from (9.2.5) that $J_n(z)$, as defined by (9.1.1), is a solution of the ordinary linear differential equation of second order

$$z^2w'' + zw' + (z^2 - n^2)w = 0. \tag{9.3.1}$$

This equation has a regular singular point at 0, and an irregular singular point at ∞. The location of the singular points, which are mathematical properties of the equation, does not depend on the fact that n is an integer. We therefore examine solutions of Bessel's differential equation

$$z^2w'' + zw' + (z^2 - \nu^2)w = 0, \tag{9.3.2}$$

where ν is a parameter which in general is complex.

The exponents at the origin are $\pm\nu$.

Since (9.3.2) only contains ν^2 we may for the moment assume that

$$-\frac{\pi}{2} < \arg \nu \leqslant \frac{\pi}{2}.$$

Accordingly (9.3.2) has a solution which behaves like z^ν in the neighbourhood of 0, and so to be consistent with (9.1.3) we define the Bessel function of the first kind and of order ν by

$$J_\nu(z) = \left(\frac{z}{2}\right)^\nu \sum_{r=0}^{\infty} \frac{(-1)^r(z/2)^{2r}}{\Gamma(\nu + r + 1)r!}. \tag{9.3.3}$$

This solution was obtained in section 1.4. It follows from (9.3.3) that $J_\nu(z)$ has a branch point at the origin if ν is not an integer. On the other hand, $z^{-\nu}J_\nu(z)$ is an entire function of z.

If ν is not an integer we may also construct a solution from the other root $\rho = -\nu$, and since this behaves like $z^{-\nu}$ near $z = 0$, it is independent of (9.3.3). Hence we can take as our second solution

$$J_{-\nu}(z) = \left(\frac{z}{2}\right)^{-\nu} \sum_{r=0}^{\infty} \frac{(-1)^r(z/2)^{2r}}{\Gamma(-\nu + r + 1)r!}, \tag{9.3.4}$$

which was also given in section 1.4.

If ν is a positive integer, by (9.1.2) the series (9.3.4) degenerates to a multiple of (9.3.3). An independent second solution must involve logarithmic terms, as was shown in section 1.6.

We note that when ν is half an odd integer the two exponents differ by an integer, but yet (9.3.3) and (9.3.4) give linearly independent solutions of equation (9.3.2). This happens, in effect, because the powers of z in both series increase in steps of 2.

9.4 Integral representation of $J_\nu(z)$

Equation (9.3.2) is not immediately suitable for solution by a contour integral of the Laplace type. If we substitute $w(z) = z^\nu u(z)$, Bessel's equation reduces to the more convenient form

$$zu'' + (2\nu + 1)u' + zu = 0, \tag{9.4.1}$$

and solutions of this are of the form

$$u(z) = \int_C e^{z\tau}(1 + \tau^2)^{\nu-1/2}\, d\tau, \tag{9.4.2}$$

for a suitable choice of contour C. It is more usual to replace τ by it, and so we have solutions of (9.3.2), namely

$$w(z) = A_\nu z^\nu \int_C e^{izt}(t^2 - 1)^{\nu-1/2}\, dt.$$

The only closed contour suitable for every ν is a figure of eight going round 1 in the negative direction and -1 in the positive direction. To make our definition explicit we take the argument of $(t^2 - 1)$ to be zero when t is real and greater than unity.

This contour C_2 was considered in section 8.10, and is shown in Fig. 8.1. We may distort the contour into small broken circles $|t \pm 1| = \delta$ and the real line segment $-1 + \delta \leqslant t \leqslant 1 - \delta$ taken twice (see Fig. 8.2) as we did in section 8.10.

If $\mathscr{R}(\nu + \frac{1}{2}) > 0$, it follows that

$$w(z) = A_\nu\{e^{i(\nu-1/2)\pi} - e^{-i(\nu-1/2)\pi}\}z^\nu \int_{-1}^{1} e^{izt}(1 - t^2)^{\nu-1/2}\, dt.$$

Since e^{izt} is an entire function, and the line of integration is finite,

$$\int_{-1}^{1} e^{izt}(1 - t^2)^{\nu-1/2}\, dt = \sum_{s=0}^{\infty} \frac{(iz)^s}{s!} \int_{-1}^{1} t^s(1 - t^2)^{\nu-1/2}\, dt.$$

Integrals like those of the coefficients in the series were considered in section 8.5. If s is odd they vanish, since the integrand is an odd function of t, while if $s = 2r$ they have the value $B(r + \frac{1}{2}, \nu + \frac{1}{2})$. Thus we obtain

$$w(z) = -2\pi i A_\nu \frac{\sqrt{\pi}2^\nu}{\Gamma(\frac{1}{2} - \nu)}\left(\frac{z}{2}\right)^\nu \sum_{r=0}^{\infty} \frac{(-1)^r(z/2)^{2r}}{\Gamma(\nu + r + 1)r!}.$$

With the figure-of-eight contour C_2

$$\mathcal{J}_\nu(z) = \frac{\Gamma(\frac{1}{2} - \nu)}{\sqrt{\pi}}\left(\frac{z}{2}\right)^\nu \frac{1}{2\pi i}\int_{C_2} e^{izt}(t^2 - 1)^{\nu-1/2}\, dt, \tag{9.4.3}$$

while when $\mathscr{R}(\nu + \frac{1}{2}) > 0$

$$\mathcal{J}_\nu(z) = \frac{1}{\sqrt{\pi}\,\Gamma(\nu + \frac{1}{2})}\left(\frac{z}{2}\right)^\nu \int_{-1}^{1} e^{izt}(1 - t^2)^{\nu-1/2}\, dt. \tag{9.4.4}$$

9.5 A second integral representation

We might alternatively look for solutions of (1.4.1) of the form

$$u(z) = \int_C \exp(tz^2) f(t)\, dt. \tag{9.5.1}$$

It was shown in illustration 2 of Chapter 2 that a solution is

$$u(z) = \int_C t^{\nu+1} \exp\left(z^2 t - \frac{1}{4t}\right) dt$$

for a suitable choice of C. It is more convenient to introduce a new variable of integration ζ, where $4t\zeta + 1 = 0$. Then

$$u(z) = B_\nu \int_{C'} \zeta^{-\nu-1} \exp(\zeta - z^2/4\zeta)\, d\zeta, \tag{9.5.2}$$

is a solution of (9.4.1) provided that

$$\Delta_{C'}\zeta^{-\nu-3} \exp(\zeta - z^2/4\zeta) = 0.$$

A suitable choice for C' is shown in Fig. 5.2(a). Since this contour is infinite, it is not immediately obvious that we may expand the exponential and integrate term by term. We notice that the finite contour obtained from C' by removing that part of the contour outside $|\zeta| = X$, where X is some arbitrary large number, gives a result which differs from (9.5.2) only in terms of order e^{-X}, so justifying the expansion. Thus

$$u(z) = B_\nu \sum_{r=0}^{\infty} \frac{(-1)^r}{r!}\left(\frac{z}{2}\right)^{2r} \int_{C'} \zeta^{-\nu-r-1} e^\zeta\, d\zeta.$$

The integrals in the coefficients are the Hankel integrals for the reciprocal of the gamma function (see equation (5.8.6.)) It follows that

$$u(z) = 2\pi i B_\nu \sum_{r=0}^{\infty} \frac{(-1)^r (z/2)^{2r}}{\Gamma(\nu + r + 1) r!}$$

and hence

$$\mathcal{J}_\nu(z) = \left(\frac{z}{2}\right)^\nu \frac{1}{2\pi i}\int_{C'} \zeta^{-\nu-1} \exp(\zeta - z^2/4\zeta)\, d\zeta. \tag{9.5.3}$$

Provided that $|z| \neq 0$, we say put $2\zeta = z\sigma$, and if $\mathscr{R}(z) > 0$, we may then distort the image contour back into C'. If, on the other hand, $\mathscr{R}(z) \leqslant 0$, we rotate the 'tail' of C' to ensure that the resulting integral converges.

When $\mathscr{R}(z) > 0$ we obtain

$$J_\nu(z) = \frac{1}{2\pi i}\int_{C'} \sigma^{-\nu-1} \exp\left\{\frac{z}{2}(\sigma - \sigma^{-1})\right\} d\sigma \tag{9.5.4}$$

which is to be compared with (9.1.4). In particular, we may take C' to be the unit circle broken at $\sigma = e^{\pm i\pi}$, together with the two sides of the negative real axis. We expand (9.5.4) to give

$$J_\nu(z) = \frac{1}{2\pi i}\Big[e^{\nu i\pi}\int_\infty^1 \rho^{-\nu-1}\exp\left\{-\frac{z}{2}\left(\rho - \frac{1}{\rho}\right)\right\} d\rho + i\int_{-\pi}^{\pi} \exp(iz\sin\theta - i\nu\theta)\, d\theta + e^{-\nu i\pi}\int_1^\infty \rho^{-\nu-1}\exp\left\{-\frac{z}{2}\left(\rho - \frac{1}{\rho}\right)\right\} d\rho\Big]$$

or

$$J_\nu(z) = \frac{1}{\pi}\int_0^\pi \cos(z\sin\theta - \nu\theta)\, d\theta - \sin\nu\pi\int_1^\infty \rho^{-\nu-1}\exp\left\{-\frac{z}{2}\left(\rho - \frac{1}{\rho}\right)\right\} d\rho. \tag{9.5.5}$$

This result should be compared with (9.1.5).

9.6 General recurrence relations

We have previously deduced recurrence relations for the functions of integral order. We can now show that they also hold for $J_\nu(z)$.

Since we can differentiate (9.5.3) under the integral sign, we derive

$$\frac{d^k}{d(z^2)^k}\{z^{-\nu}J_\nu(z)\} = \frac{(-1)^k}{2^{\nu+2k}}\frac{1}{2\pi i}\int_{C'} \zeta^{-\nu-k-1}\exp(\zeta - z^2/4\zeta)\, d\zeta$$
$$= (-\tfrac{1}{2})^k z^{-\nu-k}J_{\nu+k}(z). \tag{9.6.1}$$

In particular, take $k = 1$ and we get

$$\frac{d}{dz}\{z^{-\nu}J_\nu(z)\} = -z^{-\nu}J_{\nu+1}(z), \tag{9.6.2}$$

which agrees with (9.2.4). A special case of this is the well-known relation

$$J_0'(z) = -J_1(z). \tag{9.6.3}$$

If we differentiate (9.5.4) we obtain

$$\mathcal{J}_\nu'(z) = \frac{1}{4\pi i}\int_{C'}(\sigma^{-\nu} - \sigma^{-\nu-2})\exp\left\{\frac{z}{2}(\sigma - \sigma^{-1})\right\}d\sigma$$
$$= \tfrac{1}{2}\{\mathcal{J}_{\nu-1}(z) - \mathcal{J}_{\nu+1}(z)\}, \tag{9.6.4}$$

which is the generalization of (9.2.2).

We obtain from (9.6.2) and (9.6.4) a relation equivalent to (9.2.3), namely

$$\frac{d^k}{d(z^2)^k}\{z^\nu \mathcal{J}_\nu(z)\} = (\tfrac{1}{2})^k z^{\nu-k}\mathcal{J}_{\nu-k}(z). \tag{9.6.5}$$

This result can be further generalized for non-integral k. Let $\mathscr{R}(\nu) > 0$ and $\mathscr{R}(\lambda) > 0$ and consider

$$\frac{1}{\Gamma(\lambda)}\int_0^\zeta (\zeta - t)^{\lambda-1}t^{\nu/2}\mathcal{J}_\nu(2\sqrt{t})\,dt$$
$$= \frac{1}{\Gamma(\lambda)}\sum_{n=0}^{\infty}\frac{(-1)^n}{\Gamma(n+\nu+1)n!}\int_0^\zeta(\zeta - t)^{\nu-1}t^{\nu+n}\,dt$$
$$= \frac{1}{\Gamma(\lambda)}\sum_{n=0}^{\infty}\frac{(-1)^n B(\lambda, \nu+n+1)}{\Gamma(n+\nu+1)n!}\zeta^{\lambda+\nu+n}$$
$$= \zeta^{(\lambda+\nu)/2}\mathcal{J}_{\nu+\lambda}(2\sqrt{\zeta}). \tag{9.6.6}$$

If we put $4\zeta = z^2$ and $\lambda = k$, this is equivalent to (9.6.5), with the differentiations replaced by integrations.

We may also obtain a generalization of the other recurrence relation (9.6.1), though it takes longer to establish since we cannot integrate term by term. We notice initially that $t^{-\nu/2}\mathcal{J}_\nu(2\sqrt{t})$ is a solution of the differential equation

$$t\frac{d^2v}{dt^2} + (\nu+1)\frac{dv}{dt} + v = 0.$$

Let

$$w(\zeta) = \frac{1}{\Gamma(\lambda)}\int_\zeta^\infty (t - \zeta)^{\lambda-1}v(t)\,dt,$$

where $0 < \mathscr{R}(\lambda) < \mathscr{R}\{\tfrac{1}{2}(\nu+1)\}$ and $v(t) = t^{-\nu/2}\mathcal{J}_\nu(2\sqrt{t})$. It follows that

$$\frac{dw}{d\zeta} = \frac{1}{\Gamma(\lambda)}\int_\zeta^\infty (t-\zeta)^{\lambda-1}\frac{dv}{dt}\,dt,$$
$$\frac{d^2w}{d\zeta^2} = \frac{1}{\Gamma(\lambda)}\int_\zeta^\infty (t-\zeta)^{\lambda-1}\frac{d^2v}{dt^2}\,dt,$$

and so

$$\zeta\frac{d^2w}{d\zeta^2} + (\nu - \lambda + 1)\frac{dw}{d\zeta} + w = -\frac{1}{\Gamma(\lambda)}\int_0^\infty \frac{d}{dt}\{(t-\zeta)^\lambda v(t)\}\,dt = 0.$$

Thus $w(\zeta)$ satisfies a differential equation similar to that for $v(t)$ but with $\nu - \lambda$ replacing ν, and it follows that

$$\zeta^{-(\nu-\lambda)/2}\mathcal{J}_{\nu-\lambda}(2\sqrt{\zeta}) = \frac{1}{\Gamma(\lambda)}\int_\zeta^\infty (t-\zeta)^{\lambda-1}t^{-\nu/2}\mathcal{J}_\nu(2\sqrt{t})\,dt. \tag{9.6.7}$$

The restrictions can be relaxed a little if the line integrals are replaced by contour integrals.

9.7 The zeros of $\mathcal{J}_\nu(z)$

We shall show in the next section that when ν is real all the zeros of $\mathcal{J}_\nu(z)$ are either real or purely imaginary. Moreover, when $\nu > -1$ they are all real. In this section we prove that there are an infinity of real zeros when ν is real.

The transformation

$$w(z) = z^{-1/2}v(z)$$

reduces equation (9.3.2) to the normal form

$$\frac{d^2v}{dz^2} + \{1 - (\nu^2 - \tfrac{1}{4})z^{-2}\}v = 0. \tag{9.7.1}$$

Let us consider only positive real values of z. Then given $\varepsilon > 0$, we can find $x_0(\varepsilon, \nu)$ such that

$$\varepsilon x_0^2 > |\nu^2 - \tfrac{1}{4}|.$$

Following the ideas of section 3.3 we may 'compare' (9.7.1) with the equations

$$\frac{d^2u}{dx^2} + (1 \pm \varepsilon)u = 0$$

in the interval $x_0 < x < \infty$. We deduce that $v(x)$ oscillates more quickly than $\cos\{(1-\varepsilon)^{1/2}x + \delta\}$ and more slowly than $\cos\{(1+\varepsilon)^{1/2}x + \delta\}$.

It then follows that $v(x)$, and so also any solution $w(z)$ of (9.3.2), has an infinity of real zeros. Since we can make ε as small as we like, increasing x_0 accordingly, it follows that as $|x| \to \infty$, the spacing of the zeros is asymptotically the same as those of $\cos(x + \delta)$. Thus the separation of consecutive zeros will tend to π as x increases.

Now $z^{-\nu}\mathcal{J}_\nu(z)$ satisfies equation (9.4.1)

$$zu'' + (2\nu + 1)u' + zu = 0,$$

and with the exception possibly of 0, $u(z)$ has the same zeros as $\mathcal{J}_\nu(z)$. From (9.4.1) we see that at a zero of u', u and u'' take opposite signs.

Neither u, u' nor u'' can have a double zero, as they all satisfy second-order differential equations whose only singular point is 0. (Since a solution with a double zero vanishes with its first derivative at the zero, it follows from the uniqueness theorem for ordinary differential equations that it vanishes identically.) Thus by Rolle's theorem u'' vanishes an odd number of times between every pair of consecutive zeros of u' and hence takes opposite signs at a pair of such zeros. Therefore u takes opposite signs at consecutive zeros of u', and must vanish an odd number of times between. But Rolle's theorem also implies that u' vanishes an odd number of times between consecutive zeros of u, and hence the zeros of u and u' must interlace.

From the recurrence relations

$$u' = \frac{\mathrm{d}}{\mathrm{d}z}\{z^{-\nu}\mathcal{J}_\nu(z)\} = -z^{-\nu}\mathcal{J}_{\nu+1}(z),$$

it follows that, excepting possibly $z = 0$, the zeros of $\mathcal{J}_\nu(z)$ and $\mathcal{J}_{\nu+1}(z)$ interlace.

9.8 The orthogonality relations

The function $\mathcal{J}_\nu(\alpha t)$ satisfies the equation

$$\frac{\mathrm{d}}{\mathrm{d}t}\left\{t\frac{\mathrm{d}\mathcal{J}_\nu(\alpha t)}{\mathrm{d}t}\right\} + (\alpha^2 t - \nu^2/t)\mathcal{J}_\nu(\alpha t) = 0. \qquad (9.8.1)$$

We may have a similar equation for $\mathcal{J}_\nu(\beta t)$. We now multiply (9.8.1) by $\mathcal{J}_\nu(\beta t)$ and subtract from it the complementary equation multiplied by $\mathcal{J}_\nu(\alpha t)$ to derive

$$\frac{\mathrm{d}}{\mathrm{d}t}\left[t\left\{\mathcal{J}_\nu(\beta t)\frac{\mathrm{d}\mathcal{J}_\nu(\alpha t)}{\mathrm{d}t} - \mathcal{J}_\nu(\alpha t)\frac{\mathrm{d}\mathcal{J}_\nu(\beta t)}{\mathrm{d}t}\right\}\right] + (\alpha^2 - \beta^2)t\mathcal{J}_\nu(\alpha t)\mathcal{J}_\nu(\beta t) = 0. \qquad (9.8.2)$$

If we now assume that $\mathscr{R}(\nu) > 0$ and integrate from $t = 0$ to $t = 1$, this becomes

$$\alpha\mathcal{J}_\nu(\beta)\mathcal{J}_\nu'(\alpha) - \beta\mathcal{J}_\nu(\alpha)\mathcal{J}_\nu'(\beta) + (\alpha^2 - \beta^2)\int_0^1 t\mathcal{J}_\nu(\alpha t)\mathcal{J}_\nu(\beta t)\,\mathrm{d}t = 0. \qquad (9.8.3)$$

We know that except for the origin $\mathcal{J}_\nu(z)$ is well behaved, and so we may substitute $\beta = \alpha + \delta\alpha$ and expand each term as a Taylor series. This gives

$$\alpha\mathcal{J}_\nu(\alpha)\mathcal{J}_\nu'(\alpha) + \alpha\{\mathcal{J}_\nu'(\alpha)\}^2\delta\alpha + O\{(\delta\alpha)^2\} - \alpha\mathcal{J}_\nu(\alpha)\mathcal{J}_\nu'(\alpha)$$
$$- \{\mathcal{J}_\nu(\alpha)\mathcal{J}_\nu'(\alpha) + \alpha\mathcal{J}_\nu(\alpha)\mathcal{J}_\nu''(\alpha)\}\delta\alpha - O\{(\delta\alpha)^2\}$$
$$- 2\alpha\delta\alpha\int_0^1 t\{\mathcal{J}_\nu(\alpha t)\}^2\,\mathrm{d}t + O\{(\delta\alpha)^2\} = 0.$$

The terms of order $\delta\alpha$ vanish separately and so

$$\alpha\{\mathcal{J}_\nu'(\alpha)\}^2 - \mathcal{J}_\nu(\alpha)\mathcal{J}_\nu'(\alpha) - \alpha\mathcal{J}_\nu(\alpha)\mathcal{J}_\nu''(\alpha) - 2\alpha\int_0^1 t\{\mathcal{J}_\nu(\alpha t)\}^2\,dt = 0. \tag{9.8.4}$$

If in (9.8.3) we assume that α and β are distinct real zeros of $\mathcal{J}_\nu(z)$, we deduce the orthogonality relation

$$\int_0^1 t\mathcal{J}_\nu(\alpha t)\mathcal{J}_\nu(\beta t) = 0. \tag{9.8.5}$$

Correspondingly we deduce from (9.8.4) that when α is a real zero of $\mathcal{J}_\nu(z)$, then

$$\int_0^1 t\{\mathcal{J}_\nu(\alpha t)\}^2\,dt = \tfrac{1}{2}\{\mathcal{J}_\nu'(\alpha)\}^2 = \tfrac{1}{2}\{\mathcal{J}_{\nu+1}(\alpha)\}^2. \tag{9.8.6}$$

This provides a formal method of expanding an arbitrary function in a series of Bessel functions, the Fourier–Bessel expansion

$$f(t) = \sum_{n=1}^{\infty} a_n\mathcal{J}_\nu(\alpha_n t)$$

where the α_n are the set of positive zeros of $\mathcal{J}_\nu(z)$ and from (9.8.5) and (9.8.6)

$$\{\mathcal{J}_{\nu+1}(\alpha_n)\}^2 a_n = 2\int_0^1 tf(t)\mathcal{J}_\nu(\alpha_n t)\,dt.$$

Sufficient conditions for the function $f(t)$ to allow such an expansion are found in [15]. We have also discussed this problem in section 3.10.

When ν is real, the expansion of $\mathcal{J}_\nu(z)$ has real coefficients. It follows that if α is a complex zero of $\mathcal{J}_\nu(z)$, so also is its complex conjugate $\bar{\alpha}$.

Substituting $\beta = \bar{\alpha}$ into (9.8.3), we derive

$$(\alpha^2 - \bar{\alpha}^2)\int_0^1 t\,|\,\mathcal{J}_\nu(\alpha t)\,|^2\,dt = 0.$$

Since the integral is essentially positive,

$$\alpha^2 = \bar{\alpha}^2$$

and so α is either real or purely imaginary. Now

$$\mathcal{J}_\nu(iy) = e^{i\nu\pi/2}\left(\frac{y}{2}\right)^\nu \sum_{n=0}^{\infty} \frac{(y/2)^{2n}}{\Gamma(\nu+n+1)\,\Gamma(n+1)}$$

and for $\nu > -1$ every coefficient in the series is positive. Hence when $\nu > -1$, $\mathcal{J}_\nu(z)$ can have no pure imaginary roots, and so all its roots are real. Even when $\nu < -1$, only the first few coefficients can be negative and so at worst the pure imaginary zeros can form a finite set.

9.9 Independent solutions of the equation

We know that, unless ν is an integer, independent solutions of the Bessel equation are $J_\nu(z)$ and $J_{-\nu}(z)$, and so all solutions can be expressed in terms of them. When ν is an integer n, we have seen that $J_n(z)$ and $J_{-n}(z)$ are no longer independent and so we must derive a second solution by other means. We have seen in section 1.6 that such a second solution will include terms like z^{-n} and also terms like $J_n(z) \log z$. We shall now derive a second solution directly, rather than appeal to section 1.6.

From (9.3.3) we can see that provided $\mathscr{R}(z) > 0$ another suitable contour for a solution of the Bessel equation is the one we denote by C_3 (see Fig. 9.1(a)). Hence a solution, called the Hankel function of order ν of the first kind, is

$$H_\nu^{(1)}(z) = \frac{2\Gamma(\frac{1}{2} - \nu)}{\sqrt{\pi}}\left(\frac{z}{2}\right)^\nu \frac{1}{2\pi i}\int_{C_3} e^{izt}(t^2 - 1)^{\nu-1/2}\,dt. \qquad (9.9.1)$$

We take $\arg(t^2 - 1) = 0$ at the point on C_3 where t is real and greater than unity. When the contour C_3 is rotated until it makes an angle α

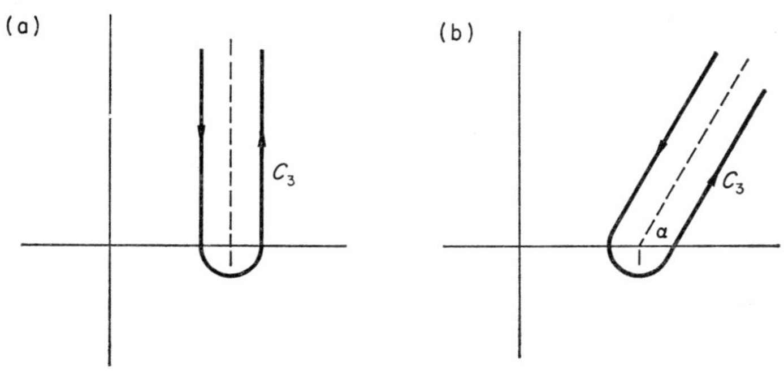

Fig. 9.1

with the positive real axis (Fig. 9.1(b)), the integral still converges provided

$$y \cos \alpha + x \sin \alpha > 0,$$

or equivalently

$$0 < \alpha + \arg z < \pi.$$

Now α can be rotated from $-\pi$ to π, and since if we choose a sequence of values of α we have correspondingly a sequence of overlapping half-planes, we can continue the function analytically into

$$-\pi < \arg z < 2\pi.$$

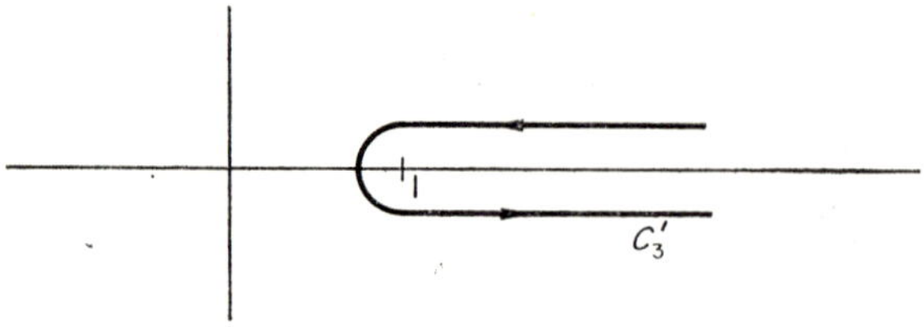

Fig. 9.2

In particular when $z = iy$, where y is real and positive, we take $\alpha = 0$ (Fig. 9.2) and so obtain

$$H_\nu^{(1)}(iy) = \frac{2\Gamma(\frac{1}{2}-\nu)}{\sqrt{\pi}} e^{i\pi\nu/2}\left(\frac{y}{2}\right)^\nu \frac{1}{2\pi i}\int_{C_3'} e^{-yt}(t^2-1)^{\nu-1/2}\,dt \quad (9.9.2)$$

which is clearly $O(e^{-y})$ as $y \to \infty$. On the other hand it is exponentially large for negative y, that is, when z is in the lower half plane.

We may similarly define the Hankel function of order ν of the second kind in $\mathscr{R}(z) > 0$ by

$$H_\nu^{(2)}(z) = \frac{2\Gamma(\frac{1}{2}-\nu)}{\sqrt{\pi}}\left(\frac{z}{2}\right)^\nu \frac{1}{2\pi i}\int_{C_4} e^{izt}(t^2-1)^{\nu-1/2}\,dt. \quad (9.9.3)$$

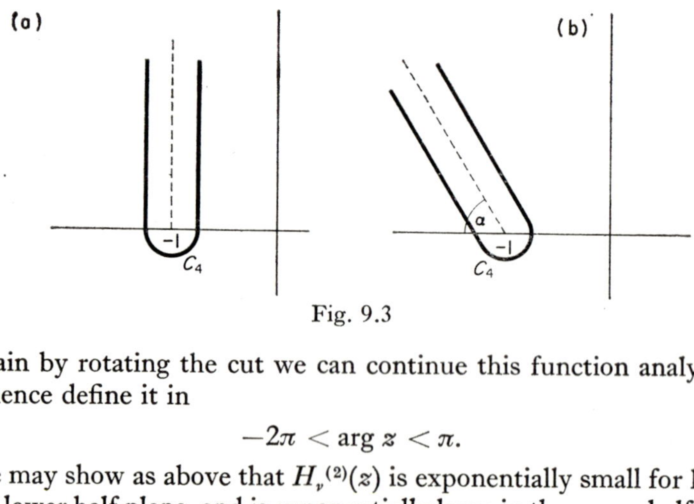

Fig. 9.3

Again by rotating the cut we can continue this function analytically and hence define it in

$$-2\pi < \arg z < \pi.$$

We may show as above that $H_\nu^{(2)}(z)$ is exponentially small for large z in the lower half plane, and is exponentially large in the upper half plane. Thus the two solutions are certainly independent.

We notice also that the composite contour $C_2 + C_3$ can be distorted into a figure of eight about the points ± 1, and so

$$\mathcal{J}_\nu(z) = \tfrac{1}{2}\{H_\nu^{(1)}(z) + H_\nu^{(2)}(z)\}. \quad (9.9.4)$$

The Bessel equation contains ν only in the form ν^2, and so we can get other integral solutions of the equation by writing $-\nu$ for ν in (9.9.1) and in (9.9.3). We see, however, that the resulting integrals are exponentially small or large in the same half planes as the original. Thus, for example,

$$H_{-\nu}^{(1)}(z) = A_\nu H_\nu^{(1)}(z) \tag{9.9.5}$$

where A_ν is a constant. We shall show later that if $|z| \gg 1$, then

$$H_\nu^{(1)}(z) \sim \left(\frac{2}{\pi z}\right)^{1/2} \exp\left\{\mathrm{i}\left(z - \frac{\nu\pi}{2} - \frac{\pi}{4}\right)\right\},$$

and hence it follows that $A_\nu = \mathrm{e}^{\mathrm{i}\nu\pi}$. Similarly we may show that

$$H_{-\nu}^{(2)}(z) = \mathrm{e}^{-\mathrm{i}\nu\pi} H_\nu^{(2)}(z). \tag{9.9.6}$$

Then corresponding to (9.9.4) we have

$$J_{-\nu}(z) = \tfrac{1}{2}\{\mathrm{e}^{\mathrm{i}\nu\pi} H_\nu^{(1)}(z) + \mathrm{e}^{-\mathrm{i}\nu\pi} H_\nu^{(2)}(z)\} \tag{9.9.7}$$

and $J_{-\nu}(z)$ is an independent second solution of the equation if ν is not an integer, but is $(-1)^n J_n(z)$ in the case $\nu = n$. From (9.8.4) and (9.8.7) it follows that

$$\mathrm{i}H_\nu^{(1)}(z) = \{J_{-\nu}(z) - J_\nu(z)\,\mathrm{e}^{-\mathrm{i}\nu\pi}\} \operatorname{cosec} \nu\pi$$

and

$$\mathrm{i}H_\nu^{(2)}(z) = \{\mathrm{e}^{\mathrm{i}\nu\pi} J_\nu(z) - J_{-\nu}(z)\} \operatorname{cosec} \nu\pi.$$

In consequence

$$\begin{aligned} Y_\nu(z) &= \frac{1}{2\mathrm{i}}\{H_\nu^{(1)}(z) - H_\nu^{(2)}(z)\} \\ &= \{J_\nu(z) \cos \nu\pi - J_{-\nu}(z)\} \operatorname{cosec} \nu\pi \end{aligned} \tag{9.9.8}$$

is a solution of the equation for all ν as it is a linear combination of $H_\nu^{(1)}(z)$ and $H_\nu^{(2)}(z)$, and in the limit $\nu \to n$ it remains finite. It therefore gives an alternative second solution of the equation, which is still defined when $\nu = n$. We call $Y_\nu(z)$ the Bessel function of the second kind.

If in (9.9.8) we let $\nu \to n$ and apply the limiting method of l'Hôpital, we get an explicit expression for $Y_n(z)$, namely

$$Y_n(z) = \frac{1}{\pi}\left\{\frac{\partial J_\nu(z)}{\partial \nu} - (-1)^n \frac{\partial J_{-\nu}(z)}{\partial \nu}\right\}_{\nu=n}. \tag{9.9.9}$$

We may differentiate the series (9.3.3) and (9.3.4) and substitute $\nu = n$, which gives

$$\begin{aligned} Y_n(z) = \frac{1}{\pi}\Bigg[\sum_{r=0}^{\infty} \frac{(-1)_r}{\{(n+r)!\}n!}\left\{2\log\left(\frac{z}{2}\right) + 2\gamma - \sum_{k=1}^{n+r} k^{-1} - \sum_{k=1}^{r} k^{-1}\right\}\left(\frac{z}{2}\right)^{2r} \\ - \sum_{r=0}^{n-1} \frac{\{(n-r-1)!\}}{r!}\left(\frac{z}{2}\right)^{2r-n}\Bigg]. \end{aligned} \tag{9.9.10}$$

An integral form for $Y_\nu(z)$ can be deduced from (9.5.5) and (9.9.8), namely

$$Y_\nu(z) = \frac{1}{\pi}\int_0^\pi \sin\{z\sin\theta - \nu\theta\}\,\mathrm{d}\theta$$
$$- \int_1^\infty \{\rho^{-\nu-1}\cos\nu\pi + \rho^{\nu-1}\}\exp\left\{-\frac{z}{2}\left(\rho - \frac{1}{\rho}\right)\right\}\mathrm{d}\rho. \quad (9.9.11)$$

From (9.2.1), (9.2.2) and (9.9.8) it is easily seen that $Y_\nu(z)$ satisfies the same recurrence relations as $J_\nu(z)$, and therefore it follows that $H_\nu^{(1)}(z)$ and $H_\nu^{(2)}(z)$ also satisfy the same recurrence relations.

The function $Y_n(z)$ was the second solution defined by Weber. Alternative solutions which are sometimes useful are

$$\begin{aligned} \mathbf{Y}_\nu(z) &= 2\pi\,\mathrm{e}^{\mathrm{i}\nu\pi}\operatorname{cosec} 2\nu\pi\{J_\nu(z)\cos\nu\pi - J_{-\nu}(z)\} \\ &= \pi\,\mathrm{e}^{\mathrm{i}\nu\pi}\sec\nu\pi\, Y_\nu(z), \end{aligned}$$

which was given by Hankel, and

$$Y^{(\nu)}(z) = \tfrac{1}{2}\mathbf{Y}_n(z) + (\log 2 - \gamma)J_\nu(z),$$

where γ is Euler's constant, which was given by Neumann.

9.10 The modified Bessel functions

Though the various expressions for Bessel functions given in previous sections give their value at almost every point of the complex plane, the functions with argument iz rather than z occur frequently in mathematical physics and so merit a notation of their own.

Changing z into iz in equation (9.3.2), we derive the modified Bessel equation

$$z^2w'' + zw' - (z^2 + \nu^2)w = 0. \quad (9.10.1)$$

The modified Bessel function of the first kind of order ν is denoted by $I_\nu(z)$ and is defined by

$$I_\nu(z) = \mathrm{e}^{-\mathrm{i}\nu\pi/2}J_\nu(\mathrm{i}z) = \left(\frac{z}{2}\right)^\nu \sum_{n=0}^\infty \frac{(z/2)^{2n}}{\Gamma(\nu+n+1)n!}. \quad (9.10.2)$$

Integral expressions for $I_\nu(z)$ can be obtained directly from the corresponding formulae for $J_\nu(z)$.

By making the same substitution in (9.2.1) and (9.2.2) we get recurrence relations for $I_\nu(z)$, namely

$$z\{I_{\nu-1}(z) - I_{\nu+1}(z)\} = 2\nu I_\nu(z), \quad (9.10.3)$$

and

$$I_{\nu-1}(z) + I_{\nu+1}(z) = 2I_\nu'(z). \quad (9.10.4)$$

As in the case of the ordinary functions, a second solution of (9.10.1) is provided by $I_{-\nu}(z)$, provided that ν is not an integer. It is again desirable to define a second solution which is valid whether ν is integral or not. It is now conventional to write the modified Bessel function of the second kind of order ν as $K_\nu(z)$, but to use a definition which is not in complete analogy with (9.9.8). In fact

$$K_\nu(z) = \frac{\pi}{2}\{I_{-\nu}(z) - I_\nu(z)\} \operatorname{cosec} \nu\pi. \tag{9.10.5}$$

Then from (9.9.4), (9.9.7), (9.10.2) and (9.10.5) we see that

$$K_\nu(z) = \frac{\pi i}{2} e^{i\nu\pi/2} H_\nu^{(1)}(iz). \tag{9.10.6}$$

An explicit expression for $K_n(z)$ is

$$\frac{1}{2}\sum_{r=0}^{n-1} \frac{(-1)^r\{(n-r-1)!\}}{r!}\left(\frac{z}{2}\right)^{2r-n} + (-1)^{n+1}\sum_{r=0}^{\infty}\frac{1}{\{(n+r)!\}r!}\left\{\log\left(\frac{z}{2}\right)+\gamma - \sum_{k=1}^{r}\frac{1}{k} - \sum_{k=1}^{n+r}\frac{1}{k}\right\}\left(\frac{z}{2}\right)^{n+2r}. \tag{9.10.7}$$

From (9.9.1) and (9.10.6) a convenient integral form if $\mathscr{R}(\nu) > 0$ and $\mathscr{R}(z) > 0$ is

$$K_\nu(z) = \frac{\sqrt{\pi}}{\Gamma(\nu+\frac{1}{2})}\left(\frac{z}{2}\right)^\nu \int_1^\infty e^{-zt}(t^2-1)^{\nu-1/2}\,dt$$
$$= \frac{\sqrt{\pi}}{\Gamma(\nu+\frac{1}{2})}\left(\frac{z}{2}\right)^\nu \int_0^\infty e^{-z\cosh\eta}\sinh^{2\nu}\eta\,d\eta. \tag{9.10.8}$$

In particular

$$K_{1/2}(z) = \left(\frac{\pi}{2z}\right)^{1/2} e^{-z}. \tag{9.10.9}$$

The recurrence relations for $K_\nu(z)$ differ slightly from those for $I_\nu(z)$ taking the forms

$$z\{K_{\nu-1}(z) - K_{\nu+1}(z)\} = -2\nu K_\nu(z) \tag{9.10.10}$$

and

$$K_{\nu-1}(z) + K_{\nu+1}(z) = -2K_\nu'(z). \tag{9.10.11}$$

9.11 The Wronskians

The Abel–Liouville theorem [7] gives immediately that any pair of independent solutions of either equation (9.3.2) or equation (9.10.1) have Wronskians that are proportional to $1/z$. The actual constants of proportionality are, however, of some interest.

Assuming that ν is not an integer, we see that near $z = 0$

$$J_\nu(z) = \left(\frac{z}{2}\right)^\nu \{\Gamma(\nu+1)\}^{-1}\{1 + O(z^2)\}$$

and

$$J_{-\nu}(z) = \left(\frac{z}{2}\right)^{-\nu} \{\Gamma(1-\nu)\}^{-1}\{1 + O(z^2)\}.$$

Hence

$$J_\nu(z)J_{-\nu}'(z) - J_\nu'(z)J_{-\nu}(z) = \frac{-2\nu}{z\Gamma(\nu+1)\,\Gamma(1-\nu)}\{1 + O(z^2)\}.$$

By the theorem the terms $O(z^2)$ must vanish identically, and so

$$J_\nu(z)J_{-\nu}'(z) - J_\nu'(z)J_{-\nu}(z) = -\frac{2\nu}{z\Gamma(\nu+1)\,\Gamma(1-\nu)} = -\frac{2\sin\nu\pi}{\pi z}. \tag{9.11.1}$$

Then from (9.9.8)

$$\begin{aligned} J_\nu(z)Y_\nu'(z) - J_\nu'(z)Y_\nu(z) &= -\{J_\nu(z)J_{-\nu}'(z) - J_\nu'(z)J_{-\nu}(z)\}\operatorname{cosec}\nu\pi \\ &= \frac{2}{\pi z}, \end{aligned} \tag{9.11.2}$$

and this holds in the limit as ν tends to an integer n, since the limit on both sides exists.

From (9.9.4) and (9.9.7) we also derive

$$\begin{aligned} J_\nu(z)J_{-\nu}'(z) - J_\nu'(z)J_{-\nu}(z) = -\frac{i}{2}\sin\nu\pi\{&H_\nu^{(1)}(z)H_\nu^{(2)\prime}(z) \\ &- H_\nu^{(1)\prime}(z)H_\nu^{(2)}(z)\}, \end{aligned}$$

and so

$$H_\nu^{(1)}(z)H_\nu^{(2)\prime}(z) - H_\nu^{(1)\prime}(z)H_\nu^{(2)}(z) = -\frac{4i}{\pi z}, \tag{9.11.3}$$

which establishes that $H_\nu^{(1)}(z)$ and $H_\nu^{(2)}(z)$ are independent solutions of the Bessel equation even if ν is an integer.

In the case of the modified Bessel functions, we have directly from (9.11.1)

$$I_\nu(z)I_{-\nu}'(z) - I_\nu'(z)I_{-\nu}(z) = -\frac{2\sin\nu\pi}{\pi z}, \tag{9.11.4}$$

and so from (9.10.5)

$$I_\nu(z)K_\nu'(z) - I_\nu'(z)K_\nu(z) = -\frac{1}{z}. \tag{9.11.5}$$

9.12 The Bessel function as a confluent hypergeometric function

The substitution $w(z) = z^\nu e^{-iz} v(\eta)$, where $\eta = 2iz$, transforms equation (9.3.2) into

$$\eta \frac{d^2 v}{d\eta^2} + \{2\nu + 1 - \eta\} \frac{dv}{d\eta} - (\nu - \tfrac{1}{2})v = 0, \qquad (9.12.1)$$

which is a confluent hypergeometric equation. Hence equation (9.3.2) has, under certain restrictions, the solutions

$$z^\nu e^{-iz} {}_1F_1\{\nu + \tfrac{1}{2}; 2\nu + 1; 2iz\} \quad \text{and} \quad z^{-\nu} e^{-iz} {}_1F_1(\tfrac{1}{2} - \nu; 1 - 2\nu; 2iz\}.$$

Both $\mathcal{J}_\nu(z)$ and $\mathcal{J}_{-\nu}(z)$ can therefore be expressed as linear combinations of these. We notice, however, that when $\mathscr{R}(\nu) > 0$, the second of these functions behaves like $z^{-\nu}$ near 0, and so cannot be contained in $\mathcal{J}_\nu(z)$. We may therefore equate coefficients of z^ν in the expansion round the origin and derive directly

$$\mathcal{J}_\nu(z) = \{\Gamma(\nu + 1)\}^{-1} \left(\frac{z}{2}\right) e^{-iz} {}_1F_1\{\nu + \tfrac{1}{2}; 2\nu + 1; 2iz\}. \quad (9.12.2)$$

By symmetry,

$$\mathcal{J}_{-\nu}(z) = \{\Gamma(1 - \nu)\}^{-1} \left(\frac{z}{2}\right)^{-\nu} e^{-iz} {}_1F_1\{\tfrac{1}{2} - \nu; 1 - 2\nu; 2iz\}. \quad (9.12.3)$$

We could clearly express the two Hankel functions as confluent hypergeometric functions using the results of section 9.9 and hence express them as Whittaker functions by the methods in Chapter 7. However, we notice directly that the solution of the Bessel equation which is exponentially small in the upper half plane is $H_\nu^{(1)}(z)$, and the Whittaker function $W_{k,m}(z)$ is the solution which is exponentially small in the positive half plane. Thus we see that

$$H_\nu^{(1)}(z) = A_\nu z^{-1/2} W_{0,\nu}(-2iz) \qquad (9.12.4)$$

for some constant A_ν, and in a similar way

$$H_\nu^{(2)}(z) = B_\nu z^{-1/2} W_{0,\nu}(+2iz). \qquad (9.12.5)$$

Now if $\mathscr{R}(\nu + \tfrac{1}{2}) > 0$ we can replace (9.9.1) by an integral along the line $t = 1 + i\tau$ from $\tau = 0$ to ∞, namely

$$H_\nu^{(1)}(z) = i\left(\frac{2}{\pi}\right)^{1/2} \frac{e^{-i(\nu-1)\pi/2}}{\Gamma(\frac{1}{2} + \nu)} z^\nu e^{iz} \int_0^\infty e^{-z t} \tau^{\nu - 1/2} \left(1 + \frac{i\tau}{2}\right)^{\nu - 1/2} d\tau,$$

or in the half plane $\mathscr{R}(z) > 0$

$$H_\nu^{(1)}(z) = i\left(\frac{2}{\pi z}\right)^{1/2} \frac{e^{-i(\nu-1/2)\pi/2}}{\Gamma(\frac{1}{2} + \nu)} e^{iz} \int_0^\infty e^{-\eta} \eta^{\nu - 1/2} \left(1 + \frac{i\eta}{2z}\right)^{\nu - 1/2} d\eta.$$

By comparison with

$$W_{0,\nu}(-2iz) = \frac{1}{\Gamma(\frac{1}{2}+\nu)} e^{iz} \int_0^\infty e^{-\eta}\eta^{\nu-1/2}\left(1+\frac{i\eta}{2z}\right)^{\nu-1/2} d\eta$$

we have

$$A_\nu = \left(\frac{2}{\pi}\right)^{1/2} e^{-i(\nu+1/2)\pi/2}. \qquad (9.12.6)$$

A similar argument yields

$$B_\nu = \left(\frac{2}{\pi}\right)^{1/2} e^{i(\nu+1/2)\pi/2}. \qquad (9.12.7)$$

9.13 Asymptotic expansions for the Bessel functions

We may use (9.12.4) and (9.12.5) combined with the results of section 7.8 to write down asymptotic expansions for the two Hankel functions. This gives

$$H_\nu^{(1)}(z) \sim \left(\frac{2}{\pi z}\right)^{1/2} \exp\left\{iz - i(\nu+\tfrac{1}{2})\frac{\pi}{2}\right\}\left\{1 + i\frac{4\nu^2-1}{1!2^3z} + (i)^2\frac{(4\nu^2-1)(4\nu^2-9)}{2!\,2^6z^2} + (i)^3\frac{(4\nu^2-1)(4\nu^2-9)(4\nu^2-25)}{3!\,2^9z^3} + \dots\right\} \qquad (9.13.1)$$

and in a similar way

$$H_\nu^{(2)}(z) \sim \left(\frac{2}{\pi z}\right)^{1/2} \exp\left\{-iz + i(\nu+\tfrac{1}{2})\frac{\pi}{2}\right\}\left\{1 - i\frac{4\nu^2-1}{1!2^3z} + (-i)^2\frac{(4\nu^2-1)(4\nu^2-9)}{2!\,2^6z^2} + (-i)^3\frac{(4\nu^2-1)(4\nu^2-9)(4\nu^2-25)}{3!\,2^9z^3} + \dots\right\}. \qquad (9.13.2)$$

Then from (9.9.4) and (9.9.8)

$$J_\nu(z) = \left(\frac{2}{\pi z}\right)^{1/2}\left[U_\nu(z)\cos\left\{z-(\nu+\tfrac{1}{2})\frac{\pi}{2}\right\} - V_\nu(z)\sin\left\{z-(\nu+\tfrac{1}{2})\frac{\pi}{2}\right\}\right] \qquad (9.13.3)$$

and

$$Y_\nu(z) = \left(\frac{2}{\pi z}\right)^{1/2}\left[U_\nu(z)\sin\left\{z-(\nu+\tfrac{1}{2})\frac{\pi}{2}\right\} + V_\nu(z)\cos\left\{z-(\nu+\tfrac{1}{2})\frac{\pi}{2}\right\}\right], \qquad (9.13.4)$$

where

$$U_\nu(z) \sim 1 - \frac{(4\nu^2 - 1)(4\nu^2 - 9)}{2!\,2^6 z^2} + \frac{(4\nu^2 - 1)(4\nu^2 - 9)(4\nu^2 - 25)(4\nu^2 - 49)}{4!\,2^{12} z^4} - \ldots \tag{9.13.5}$$

and

$$V_\nu(z) \sim \frac{(4\nu^2 - 1)}{1!\,2^3 z} - \frac{(4\nu^2 - 1)(4\nu^2 - 9)(4\nu^2 - 25)}{3!\,2^9 z^3} + \ldots \tag{9.13.6}$$

In a similar way we can derive asymptotic expressions for the modified functions

$$I_\nu(z) = (2\pi z)^{-1/2}\{\mathrm{e}^z W_\nu(z) + \varepsilon_\nu\, \mathrm{e}^{-z} X_\nu(z)\} \tag{9.13.7}$$

and

$$K_\nu(z) = \left(\frac{\pi}{2z}\right)^{1/2} \mathrm{e}^{-z} X_\nu(z), \tag{9.13.8}$$

where

$$W_\nu(z) \sim 1 - \frac{(4\nu^2 - 1)}{1!\,2^3 z} + \frac{(4\nu^2 - 1)(4\nu^2 - 9)}{2!\,2^6 z^2} - \frac{(4\nu^2 - 1)(4\nu^2 - 9)(4\nu^2 - 25)}{3!\,2^9 z^3} + \ldots \tag{9.13.9}$$

$$X_\nu(z) \sim 1 + \frac{(4\nu^2 - 1)}{1!2^3 z} + \frac{(4\nu^2 - 1)(4\nu^2 - 9)}{2!\,2^6 z^2} + \frac{(4\nu^2 - 1)(4\nu^2 - 9)(4\nu^2 - 25)}{3!\,2^9 z^3} + \ldots \tag{9.13.10}$$

and ε_ν is $\exp\{\pm(\nu + \frac{1}{2})\pi \mathrm{i}\}$, the sign being chosen to agree with the sign of $\arg z$.

We notice that when ν is half an odd integer all the series above terminate, and $U_\nu(z)$, $V_\nu(z)$, etc., are defined exactly by the series.

9.14 Addition formulae

We may formally derive a simple addition formula for $J_\nu(z)$ from the integral (9.5.4). Suppose $\mathscr{R}(z) > 0$, $\mathscr{R}(\zeta) > 0$, then

$$J_\nu(z + \zeta) = \frac{1}{2\pi \mathrm{i}} \int_{C'} \sigma^{-\nu-1} \exp\left\{\frac{z}{2}(\sigma - \sigma^{-1})\right\} \exp\left\{\frac{\zeta}{2}(\sigma - \sigma^{-1})\right\} \mathrm{d}\sigma. \tag{9.14.1}$$

We replace C' by C_*', the finite contour obtained from C' by removing all of it which lies outside the circle $|\sigma| = X$, where X is an arbitrarily large positive number. Then we can expand $\exp\{\xi(\sigma - \sigma^{-1})/2\}$ in a Laurent series using (9.1.1), and this will converge uniformly on C_*'. Hence

$$\int_{C_*'} \sigma^{-\nu-1} \exp\left\{\frac{(z+\zeta)}{2}(\sigma - \sigma^{-1})\right\} d\sigma = \sum_{n=-\infty}^{\infty} J_n(\zeta) \int_{C_*'} \sigma^{n-\nu-1} \times \exp\left\{\frac{z}{2}(\sigma - \sigma^{-1})\right\} d\sigma.$$

The difference between the left-hand side of the last integral and the right-hand side of (9.14.1) is $O[\exp\{-X\mathscr{R}(z+\zeta)\}]$, and so we can let $X \to \infty$.

Thus

$$J_\nu(z+\zeta) = \sum_{n=-\infty}^{\infty} J_{\nu-n}(z) J_n(\zeta). \tag{9.14.2}$$

If we use Stirling's approximation in the Bessel series it can be seen that for fixed z, as $|\nu|$ gets very large,

$$J_\nu(z) \sim (2\pi)^{-1/2} \exp\{\nu + \nu \log(z/2) - (\nu + \tfrac{1}{2}) \log \nu\},$$

and so there is no difficulty with the convergence of (9.14.2).

There is a more general addition theorem which includes (9.14.2) as a special case. In this section we shall derive it only for the two cases ν an integer and ν half an odd integer, which are those which appear most frequently in mathematical physics. We also assume that the function has a real argument.

Consider the Helmholtz partial differential equation

$$\nabla^2\phi + \lambda^2\phi = 0.$$

Single-valued solutions of this equation in polar coordinates are

$$\phi = A J_m(\lambda r) \exp(im\theta) \tag{9.14.3}$$

where m is any integer. The equation is unchanged when the origin is shifted, and hence if we do this, a solution of the form (9.14.3) in the new coordinate system can be expressed as a sum of terms in the old coordinates, and hence

$$J_n(\lambda r') \exp(in\theta') = \sum_{m=-\infty}^{\infty} A_{n,m} J_m(\lambda r) \exp(im\theta). \tag{9.14.4}$$

From Fig. 9.4 we have

$$r'^2 = r^2 - 2ar\cos\theta + a^2.$$

In the limit $\theta \to \pi$, $\theta' \to \pi$, $r' \to r + a$, and so

$$J_n\{\lambda(r+a)\} = (-1)^n \sum_{m=-\infty}^{\infty} (-1)^m A_{n,} \, J_m(\lambda r).$$

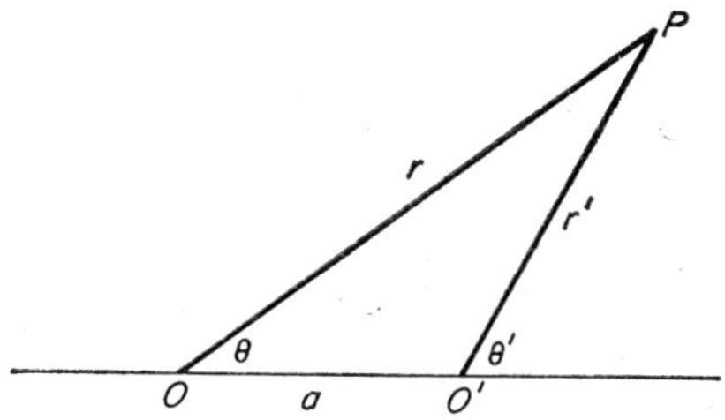

Fig. 9.4

Comparing this with (9.14.2), we have

$$A_{n,m} = (-1)^{m+n} J_{n-m}(\lambda a).$$

Hence

$$J_n(\lambda r') \exp(in\theta') = \sum_{m=-\infty}^{\infty} (-1)^{m+n} J_{n-m}(\lambda a) J_m(\lambda r) \exp(im\theta). \quad (9.14.5)$$

This may be separated into its real and imaginary parts.

We may show similarly that

$$Y_n(\lambda r') \exp(in\theta') = \sum_{m=-\infty}^{\infty} (-1)^{m+n} Y_{n-m}(\lambda a) J_m(\lambda r) \exp(im\theta), \quad (9.14.6)$$

and similar formulae can be found for $I_n(\lambda r')$ and $K_n(\lambda r')$.

In the particular case $n = 0$, (9.14.5) can be written as

$$J_0(\lambda r') = J_0(\lambda a) J_0(\lambda r) + 2 \sum_{m=1}^{\infty} J_m(\lambda a) J_m(\lambda r) \cos m\theta. \quad (9.14.7)$$

Let us adopt the usual convention of writing $\mu = \cos\theta$, when $\cos m\theta$ is the mth Chebyshev polynomial $T_m(\mu)$. Thus if both sides are differentiated with respect to μ

$$-\frac{ar}{r'} \frac{d}{dr'} J_0(\lambda r') = 2 \sum_{m=1}^{\infty} J_m(\lambda a) J_m(\lambda r) T_m'(\mu)$$

or

$$\frac{1}{r'} J_1(\lambda r') = \frac{2}{\lambda} \sum_{m=1}^{\infty} \frac{J_m(\lambda a)}{a} \frac{J_m(\lambda r)}{r} T_m'(\mu). \quad (9.14.8)$$

Similarly, n-fold differentiation yields

$$r'^{-n} J_n(\lambda r') = 2\lambda^{-n} \sum_{m=0}^{\infty} a^{-n} J_{n+m}(\lambda a) r^{-n} J_{n+m}(\lambda r) T_{n+m}{}^{(n)}(\mu). \quad (9.14.9)$$

Once again we can deduce analogous formulae for the other principal functions, namely

$$r'^{-n}Y_n(\lambda r') = 2\lambda^{-n}\sum_{m=0}^{\infty} a^{-n}Y_{n+m}(\lambda a)r^{-n}J_{n+m}(\lambda r)T_{n+m}{}^{(n)}(\mu), \tag{9.14.10}$$

$$r'^{-n}I_n(\lambda r') = 2\lambda^{-n}\sum_{m=0}^{\infty} (-1)^m a^{-n}I_{n+m}(\lambda a)r^{-n}I_{n+m}(\lambda r)T_{n+m}{}^{(n)}(\mu) \tag{9.14.11}$$

and

$$r'^{-n}K_n(\lambda r') = 2\lambda^{-n}\sum_{m=0}^{\infty} a^{-n}I_{n+m}(\lambda a)r^{-n}K_{n+m}(\lambda r)T_{n+m}{}^{(n)}(\mu), \tag{9.14.12}$$

where in the last formula it is assumed that $r > a$. Since the left-hand side is symmetric in r and a, the labelling of these is irrelevant and there is no loss of generality in taking $r > a$.

If we revert to the Helmholtz equation, and solve it in spherical polar coordinates, a solution with axial symmetry is

$$R^{-1/2}J_{n+1/2}(\lambda R)P_n(\cos\theta) \tag{9.14.13}$$

where θ is now the polar angle. We may again shift the origin and define new coordinates by

$$R'^2 = R^2 - 2aR\cos\theta + a^2 \quad\text{and}\quad \theta' = \theta + \sin^{-1}(a\sin\theta/R).$$

Then by a similar argument there exist relations of the form

$$R'^{-1/2}J_{n+1/2}(\lambda R')P_n(\cos\theta') = \sum_{m=0}^{\infty} B_{mn}R^{-1/2}J_{m+1/2}(\lambda R)P_m(\cos\theta).$$

In the particular case $n = 0$, the symmetry of the left-hand side in R and a implies that

$$\frac{\sin\lambda R'}{R'} = \sum_{m=0}^{\infty} b_m(Ra)^{-1/2}J_{m+1/2}(\lambda a)J_{m+1/2}(\lambda R)P_m(\cos\theta)$$

and it is clear dimensionally that b_m depends neither on a nor on λ. Differentiation with respect to λ yields

$$\cos\lambda R' = \sum_{m=0}^{\infty} b_m\left\{\left(\frac{a}{R}\right)^{1/2} J_{m+1/2}'(\lambda a)J_{m+1/2}(\lambda R) + \left(\frac{R}{a}\right)^{1/2} J_{m+1/2}(\lambda a)J_{m+1/2}'(\lambda R)\right\}P_m(\cos\theta),$$

and using the recurrence relation (9.6.4) we see that the right-hand side is

$$\lambda(aR)^{1/2}\sum_{m=0}^{\infty}\frac{b_m}{2m+1}\{J_{m-1/2}(\lambda a)J_{m-1/2}(\lambda R) - J_{m+3/2}(\lambda a)J_{m+3/2}(\lambda R)\} \times P_m(\cos\theta).$$

When $\theta = 0$ or $\theta = \pi$, $\cos \lambda R'$ has the simple expansion $\cos \lambda R \cos \lambda a \pm \sin \lambda R \sin \lambda a$, and so the successive terms in the series must cancel each other, leaving only terms like $\mathcal{J}_{-1/2}(\lambda R)\mathcal{J}_{-1/2}(\lambda a)$ and $\mathcal{J}_{1/2}(\lambda R)\mathcal{J}_{1/2}(\lambda a)$.

Then $b_m = (m + \frac{1}{2})\pi$ and so

$$\frac{\sin \lambda R'}{R'} = \pi \sum_{m=0}^{\infty} (m + \tfrac{1}{2})(Ra)^{-1/2}\mathcal{J}_{m+1/2}(\lambda a)\mathcal{J}_{m+1/2}(\lambda R)P_m(\cos \theta). \quad (9.14.14)$$

In the case $n = -1$ we deduce similarly that

$$\frac{\cos \lambda R'}{R'} = \sum_{m=0}^{\infty} B_{m,-1/2}R^{-1/2}\mathcal{J}_{m+1/2}(\lambda R)P_m(\cos \theta).$$

The limit of the left-hand side as $\lambda \to 0$ is $1/R'$, and if $a > R$ this can be expanded to give

$$\frac{1}{a}\sum_{m=0}^{\infty} \left(\frac{R}{a}\right)^m P_m(\cos \theta).$$

Thus

$$\frac{\cos \lambda R'}{R'} = \pi \sum_{m=0}^{\infty} (-1)^m(m + \tfrac{1}{2})(Ra)^{-1/2}\mathcal{J}_{-m-1/2}(\lambda a)\mathcal{J}_{m+1/2}(\lambda R)P_m(\cos \theta). \quad (9.14.15)$$

When $a < R$ the roles of the two constants are reversed. Let us assume that $a > R$, replace λ by $\mathrm{i}\lambda$ in both (9.14.14) and (9.14.15) and add $\mathrm{i}\lambda$ times the former to the latter. The result is

$$\frac{\exp(-\lambda R')}{R'} = \sum_{m=0}^{\infty} (2m + 1)(Ra)^{-1/2}K_{m+1/2}(\lambda a)I_{m+1/2}(\lambda R)P_m(\cos \theta). \quad (9.14.16)$$

Exercises

1. Use the generating function to show that

$$\exp(\mathrm{i}z \cos \theta) = \mathcal{J}_0(z) + 2\sum_{n=1}^{\infty} e^{n\mathrm{i}\pi}\mathcal{J}_n(z) \cos n\theta.$$

Deduce that (a) $1 = \mathcal{J}_0(z) + 2\sum_{n=1}^{\infty} \mathcal{J}_{2n}(z)$,

(b) $z(z^2 + 1) = 2\sum_{n=0}^{\infty} (2n + 1)^3\mathcal{J}_{2n+1}(z)$.

2. Show that if $a > 0$,

(a) $$\int_0^\infty \mathcal{J}_0(t)\frac{\sin at}{t}\,dt = \begin{cases}\sin^{-1} a, & 0 \leqslant a \leqslant 1\\ \pi/2, & a > 1;\end{cases}$$

(b) $$\int_0^\infty \mathcal{J}_0(t)\cos at\,dt = \begin{cases}(1-a^2)^{-1/2}, & 0 \leqslant a < 1\\ \infty, & a = 1\\ 0, & a > 1.\end{cases}$$

3. Find closed expressions for $\mathcal{J}_{1/2}(z)$, $\mathcal{J}_{-1/2}(z)$, $\mathcal{J}_{3/2}(z)$, and $\mathcal{J}_{-3/2}(z)$.

4. Show that if $\mathscr{R}(\nu + \frac{1}{2}) > 0$,

$$\left(\frac{2}{\pi}\right)^{1/2}\int_0^{\pi/2} \mathcal{J}_\nu(z\sin\theta)\sin^{\nu+1}\theta\,d\theta = z^{-1/2}\mathcal{J}_{\nu+1/2}(z).$$

5. Show that if $\mathscr{R}(\mu + \nu) > 0$,

$$\int_0^\infty e^{-st}t^{\mu-1}\mathcal{J}_\nu(t)\,dt = \frac{2^{\mu-1}}{\sqrt{\pi}}\frac{\Gamma\{(\mu+\nu)/2\}\,\Gamma\{(\mu+\nu+1)/2\}}{\Gamma(\nu+1)}$$
$$\times\, s^{-\mu-\nu}{}_2F_1\left(\frac{\mu+\nu}{2}, \frac{\mu+\nu+1}{2}; \nu+1; -\frac{1}{s^2}\right).$$

Deduce that

(a) If $\mathscr{R}(\mu) < \frac{3}{2}$, $$\int_0^\infty t^{\mu-1}\mathcal{J}_\nu(t)\,dt = \frac{2^{\mu-1}\,\Gamma\{(\mu+\nu)/2\}}{\Gamma(\nu/2-\mu/2+1)};$$

(b) If $a \geqslant 0$, $$\int_0^\infty \mathcal{J}_1(t)\frac{\sin at}{t}\,dt = \begin{cases}a-(a^2-1)^{1/2}, & a > 1\\ 1, & a = 1\\ a, & 0 \leqslant a < 1;\end{cases}$$

(c) If $\mathscr{R}(\nu+\frac{1}{2}) > 0$, $$\int_0^\infty e^{-st}t^\nu\mathcal{J}_\nu(t) = \frac{\Gamma(\nu+\frac{1}{2})}{\sqrt{\pi}(s^2+1)^{\nu+1/2}}.$$

6. Show that if n is a positive integer and $0 < \mu < 1$,

$$\int_0^\infty e^{-\mu t}t^n\mathcal{J}_0\{(1-\mu^2)^{1/2}t\}\,dt = n!P_n(\mu).$$

7. Defining $\mathcal{J}_\nu(t)$ for $t > 0$ and $\mathscr{R}(\nu+1) > 0$ by the integral

$$\mathcal{J}_\nu(t) = \frac{1}{2\pi i}\int_{C'} \sigma^{-\nu-1}\exp\left\{\frac{t}{2}(\sigma-\sigma^{-1})\right\}d\sigma,$$

where C' is as in Fig. 5.2(a), show that

$$\int_0^\infty e^{-st}\mathcal{J}_\nu(t)\,dt = (s^2+1)^{-1/2}\{(s^2+1)^{1/2}-s\}^\nu.$$

8. Show that

$$\mathcal{J}_\mu(z)\mathcal{J}_\nu(z) = \sum_{n=0}^\infty \frac{(-1)^n\,\Gamma(\mu+\nu+2n+1)(z/2)^{\mu+\nu+2n}}{\Gamma(\mu+n+1)\,\Gamma(\nu+n+1)\,\Gamma(\mu+\nu+n+1)n!}.$$

[*Hint:* prove that if $(\nu)_n = \nu(\nu-1)\ldots(\nu-n+1)$, then

$$(\mu+\nu)_n = \sum_{r=0}^n \binom{n}{r}(\mu)_{n-r}(\nu)_r.]$$

Deduce that if $\mathscr{R}(\nu + \frac{1}{2}) > 0$,

$$\int_0^{\pi/2} J_{2\nu}\{2z \cos \theta\}\, d\theta = \frac{\pi}{2}\{J_\nu(z)\}^2.$$

9. Show that if $Z_\nu(\zeta) = c_1 J_\nu(\zeta) + c_2 Y_\nu(\zeta)$, where c_1 and c_2 are any pair of arbitrary constants, then

$$w(z) = z^\alpha Z_\nu(\beta z^\gamma)$$

satisfies the differential equation

$$w'' + \frac{1 - 2\alpha}{z} w' + \left\{(b\gamma z^{\gamma-1})^2 + \frac{\alpha^2 - \nu^2\gamma^2}{z^2}\right\} w = 0.$$

Hence or otherwise find a pair of independent solutions of the differential equations

(a) $w'' + z^\nu w = 0$;

(b) $zw'' + (1 - \nu)w' + w = 0$.

10. Show that if $\mathscr{R}(\mu + \nu) > 0$, $|\arg s| < \pi/4$,

$$\int_0^\infty t^{\mu-1} J_\nu(t) \exp(-s^2t^2)\, dt = \frac{\Gamma\{(\nu+\mu)/2\}}{2^{\nu+1}s^{\mu+\nu}}\, {}_1F_1\left(\frac{\mu+\nu}{2}; \nu+1; -\frac{1}{4s^2}\right).$$

Hence or otherwise deduce that

(a) $\displaystyle\int_0^\infty t^{\nu+1} J_\nu(t) \exp(-s^2t^2)\, dt = (2s^2)^{-1-\nu} \exp(-1/4s^2)$;

(b) $\displaystyle\int_0^\infty J_{2\nu}(t) \exp(-s^2t^2)\, dt = \frac{\sqrt{\pi}}{2s} I_\nu(1/8s^2) \exp(-1/8s^2)$.

11. If $\Phi(\xi, \eta, \zeta) = (a_1 \cos \zeta + a_2 \sin \zeta) \sinh \xi \sin \eta + a_3 \cosh \xi \cos \eta$, where a_1, a_2, a_3 are three real constants, show that

$$\int_0^\pi d\eta \int_{-\pi}^{\pi} \frac{\partial \Phi}{\partial \xi} \exp\{i\Phi(\xi, \eta, \zeta)\} \sin \eta\, d\zeta = \left(\frac{2}{\pi}\right)^{1/2} \frac{|a|^2}{X^{3/2}} J_{3/2}(X)$$

where $|a|^2 = a_1^2 + a_2^2 + a_3^2$, and $X^2 = (a_1^2 + a_2^2) \sinh^2 \xi + a_3^2 \cosh^2 \xi$.

12. If $F(t)$ is of bounded total variation on $[a, b]$ where $0 \leqslant a < b < \infty$, and λ is a real positive constant, show that if $\mathscr{R}(\nu) > 0$ for large λ,

$$\int_a^b F(t) J_\nu(\lambda t)\, dt = O\left(\frac{1}{\lambda}\right).$$

13. Show that if $\nu > 0$, and λ and a are real and positive, then

$$\lim_{\lambda \to \infty} \lambda \int_0^a \frac{t}{t^2 - a^2}\{t J_\nu'(t\lambda) J_\nu(a\lambda) - a J_\nu(t\lambda) J_\nu'(a\lambda)\}\, dt = \tfrac{1}{2}$$

and

$$\lim_{\lambda \to \infty} \lambda \int_a^\infty \frac{t}{t^2 - a^2}\{t J_\nu'(t\lambda) J_2(a\lambda) - a J_\nu(t\lambda) J_\nu'(a\lambda)\}\, dt = \tfrac{1}{2}.$$

14. Deduce from exercises 12 and 13 the Hankel transform formula, that if $f(t)$ is of bounded variation on every sub-interval of $[0, \infty)$, has only a finite number of discontinuities in $[0, \infty)$, and for every t in $0 < t < \infty$

$$\int_0^t \tau^{\nu+1} \,|\, f(\tau) \,|\, d\tau < \infty \quad \text{and} \quad \int_t^\infty \tau^{1/2} \,|\, f(\tau) \,|\, d\tau < \infty,$$

then

$$\tfrac{1}{2}\{f(t+0) + f(t-0)\} = \lim_{\lambda\to\infty} \int_0^\lambda s\mathcal{J}_\nu(st)\left\{\int_0^\infty \tau f(\tau)\mathcal{J}_\nu(s\tau)\, d\tau\right\} ds.$$

15. Show that if $0 \leqslant t \leqslant 1$,

(a) $$1 = 2\sum_{k=1}^{\infty} \mathcal{J}_0(\alpha_k t) \Big/ \{\alpha_k \mathcal{J}_1(\alpha_k)\},$$

(b) $$t^2 + 1 = \sum_{k=1}^{\infty} \mathcal{J}_0(\alpha_k t)/\{\alpha_k{}^3 \mathcal{J}_1(\alpha_k)\},$$

where α_k, $k = 1, 2, 3, \ldots$ are the zeros of $\mathcal{J}_0(z)$.

16. Show that for $0 \leqslant t \leqslant 1$ and $\alpha > 0$,

(a) $$e^{-\alpha t} = \alpha \int_0^\infty \frac{s\mathcal{J}_0(st)\, ds}{(s^2 + \alpha^2)^{3/2}};$$

(b) $$f(t) = \alpha \int_0^\infty \mathcal{J}_1(\alpha s)\mathcal{J}_0(ts)\, ds,$$

where $f(t)$ is 1 if $0 \leqslant t \leqslant \alpha$, and is zero if $t > \alpha$.

17. If $x > 0$, show that

$$H_0^{(1)}(x) = \frac{2}{\pi i}\int_0^\infty \exp\,(ix \cosh \theta)\, d\theta$$

and so deduce that

$$\mathcal{J}_0(x) = \frac{2}{\pi}\int_0^\infty \sin\,(x \cosh \theta)\, d\theta, \qquad Y_0(x) = -\frac{2}{\pi}\int_0^\infty \cos\,(x \cosh \theta)\, d\theta.$$

18. Show that alternative forms for the integrals for $H_\nu^{(1)}(z)$ and $H_\nu^{(2)}(z)$ for $\mathcal{R}(z) > 0$ are

$$H_\nu^{(1)}(z) = -\frac{1}{\pi}\int_{L_1} \exp\,(-iz \sin t + i\nu t)\, dt$$

$$H_\nu^{(2)}(z) = -\frac{1}{\pi}\int_{L_2} \exp\,(-iz \sin t + i\nu t)\, dt$$

where L_1 and L_2 are shown in Figs. 9.5 and 9.6.

19. Show that if $\mathcal{R}(z) > 0$,

$$K_\nu(z) = \tfrac{1}{2}\int_{-\infty}^\infty \exp\,(-z \cosh t - \nu t)\, dt.$$

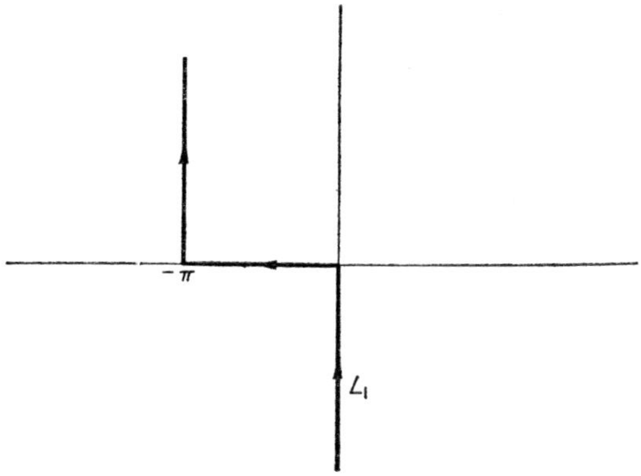

Fig. 9.5

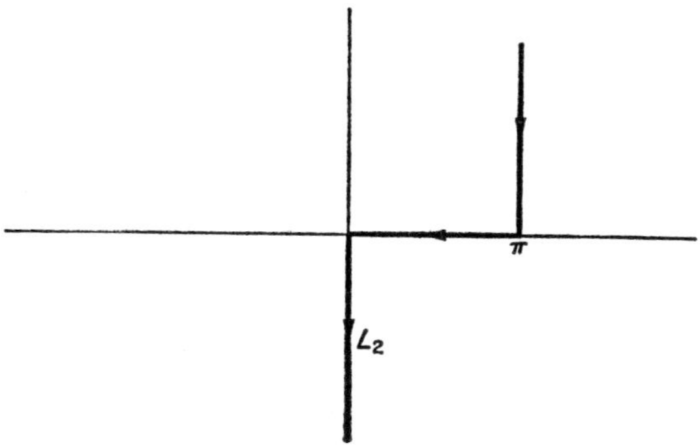

Fig. 9.6

20. Show that if $-1 < \mathscr{R}(\nu) < \frac{3}{2}$, and $a > 0$,

$$\int_0^\infty \frac{t^{\nu+1}\mathcal{J}_\nu(t)\,\mathrm{d}t}{t^2 + a^2} = a^\nu K_\nu(a).$$

21. Show that if $\mathscr{R}(s) > 1, \mathscr{R}(\nu) > 0$,

$$\int_0^\infty \mathrm{e}^{-st} t^{-1/2} I_{\nu+1/2}(t)\,\mathrm{d}t = \left(\frac{2}{\pi}\right)^{1/2} Q_\nu(s).$$

22. Show that

$$\mathcal{J}_0(2z) = \{\mathcal{J}_0(z)\}^2 + 2\sum_{n=1}^{\infty} \{\mathcal{J}_n(z)\}^2.$$

23. If $r'^2 = r^2 - 2ar\cos\theta + a^2$, show that if m is a positive integer,

(a) $\displaystyle\int_{-\pi}^{\pi} \mathcal{J}_0(\lambda r') \cos m\theta\,\mathrm{d}\theta = \varepsilon_m \pi \mathcal{J}_m(\lambda r)\mathcal{J}_m(\lambda a)$

where $\varepsilon_m = \frac{1}{2}$ if $m = 0$ and is 1 if $m \geqslant 1$;

(b) $$\int_0^\pi \frac{\sin \lambda r'}{r'} P_m(\cos\theta) \sin\theta \, d\theta = \frac{\pi}{(ar)^{1/2}} J_{m+1/2}(\lambda a) J_{m+1/2}(\lambda r);$$

and

(c) $$\int_0^\pi \frac{e^{-\lambda r'}}{r'} P_m(\cos\theta) \sin\theta \, d\theta = \frac{\pi}{(ar)^{1/2}} K_{m+1/2}(\lambda\alpha) I_{m+1/2}(\lambda\beta)$$

where if $a \geqslant r$, then $\alpha = a$, $\beta = r$, but if $r > a$, then $\alpha = r$ and $\beta = a$.

24. In the study of the motion of the earth round the sun, Lagrange showed that the mean anomaly M and the eccentric anomaly E were connected by the 'equation of time'

$$M = E - \varepsilon \sin E,$$

where ε is the eccentricity of the elliptic orbit. Show that if this equation is 'solved' in the form

$$E = M + \sum_{n=1}^{\infty} A_n \sin nM,$$

then $nA_n = 2J_n(n\varepsilon)$.

25. The vertical displacement of a circular membrane which vibrates with small amplitude satisfies the equation

$$\frac{\partial^2 y}{\partial t^2} = c^2 \left\{ \frac{\partial^2 y}{\partial r^2} + \frac{1}{r}\frac{\partial y}{\partial r} + \frac{1}{r^2}\frac{\partial^2 y}{\partial \theta^2} \right\}$$

where (r, θ) are polar coordinates of a point on the membrane. If the edge of the membrane is clamped along the circle $r = a$, and the initial displacement of the membrane is symmetric with respect to the centre of the circle, show that at a later instant the displacement can be represented in the form

$$y(r, t) = \sum_{k=1}^{\infty} A_k J_0\left(\alpha_k \frac{r}{a}\right) \cos\left(\frac{c}{a}\alpha_k t + \beta_k\right)$$

where α_k are the roots of $J_0(z) = 0$, and A_k and β_k are constants.

Answers

3. $\left(\frac{2}{\pi z}\right)^{1/2} \sin z$; $\left(\frac{2}{\pi z}\right)^{1/2} \cos z$; $\left(\frac{2}{\pi z}\right)^{1/2}\left(\frac{\sin z}{z} - \cos z\right)$;

$-\left(\frac{2}{\pi z}\right)^{1/2}\left(\frac{\sin z}{z} + \cos z\right)$.

9. (a) $z^{1/2} J_{\pm\mu}\left(\frac{2}{\mu} z^{\mu/2}\right)$ where $(\nu + 2)\mu = 1$.

(b) $z^{1/2} J_\nu(2z^{1/2})$ and $z^{1/2} Y_\nu(2z^{1/2})$.

CHAPTER 10

Laguerre Polynomials

10.1 The generating function

The Laguerre polynomials $L_n^{(\alpha)}(z)$ for non-negative integers n and $\mathscr{R}(\alpha)$ not a negative integer are defined by the generating function

$$G(z, t) \equiv (1-t)^{-1-\alpha}\, \mathrm{e}^{-zt/(1-t)} = \sum_{n=0}^{\infty} L_n^{(\alpha)}(z) t^n, \qquad |t| < 1. \qquad (10.1.1)$$

We have

$$\begin{aligned} G(z, t) &= \sum_{r=0}^{\infty} (-1)^r (zt)^r (1-t)^{-1-\alpha-r}/r! \\ &= \sum_{r=0}^{\infty} \left\{ \frac{(-1)^r (zt)^r}{r!} \sum_{n=0}^{\infty} \frac{\Gamma(n+r+\alpha+1) t^n}{n!\, \Gamma(r+\alpha+1)} \right\} \\ &= \sum_{r=0}^{\infty} \sum_{n=0}^{\infty} \frac{(-1)^r\, \Gamma(n+r+\alpha+1) z^r t^{n+r}}{r!\, n!\, \Gamma(r+\alpha+1)} \\ &= \sum_{n=0}^{\infty} \sum_{r=0}^{n} \frac{(-1)^r\, \Gamma(n+\alpha+1) z^r t^n}{r!\, (n-r)!\, \Gamma(r+\alpha+1)}, \end{aligned}$$

and so

$$L_n^{(\alpha)}(z) = \Gamma(n+\alpha+1) \sum_{r=0}^{n} \frac{(-1)^r z^r}{r!\, (n-r)!\, \Gamma(r+\alpha+1)} \qquad (10.1.2)$$

$$= \frac{\Gamma(n+\alpha+1)}{\Gamma(\alpha+1) n!}\, {}_1F_1(-n; 1+\alpha; z). \qquad (10.1.3)$$

For $\alpha = 0$, it is customary to denote the Laguerre polynomial by $L_n(z)$, and we have

$$L_n(z) = n! \sum_{r=0}^{n} \frac{(-1)^r z^r}{(r!)^2 (n-r)!} = \sum_{r=0}^{n} (-1)^r \binom{n}{r} \frac{z_r}{r!} = {}_1F_1(-n; 1; z).$$

We verify that

$$L_0^{(\alpha)}(z) = 1; \quad L_1^{(\alpha)}(z) = 1 + \alpha - z;$$
$$L_2^{(\alpha)}(z) = \tfrac{1}{2}(1 + \alpha)(2 + \alpha) - (2 + \alpha)z + \tfrac{1}{2}z^2;$$
$$L_n^{(\alpha)}(0) = \Gamma(n + \alpha + 1)/\{n!\, \Gamma(\alpha + 1)\}.$$

We can obtain another generating function which involves the Bessel function. We have

$$\begin{aligned} \mathrm{e}^t(zt)^{-\alpha/2}\mathcal{J}_\alpha\{2(zt)^{1/2}\} &= \mathrm{e}^t(zt)^{-\alpha/2}\sum_{r=0}^{\infty}\frac{(-1)^r(zt)^{\alpha/2+r}}{r!\,\Gamma(r+\alpha+1)} \\ &= \left(\sum_{n=0}^{\infty}\frac{t^n}{n!}\right)\left\{\sum_{r=0}^{\infty}\frac{(-1)^r(zt)^r}{r!\,\Gamma(r+\alpha+1)}\right\} \\ &= \sum_{n=0}^{\infty}\sum_{r=0}^{n}\frac{t^{n-r}(-1)^r(zt)^r}{(n-r)!\,r!\,\Gamma(r+\alpha+1)} \\ &= \sum_{n=0}^{\infty}\left\{t^n\sum_{r=0}^{n}\frac{(-1)^r z^r}{r!(n-r)!\,\Gamma(r+\alpha+1)}\right\}. \end{aligned}$$

That is,

$$\mathrm{e}^t(zt)^{-\alpha/2}\mathcal{J}_\alpha\{2(zt)^{1/2}\} = \sum_{n=0}^{\infty}\frac{t^n L_n^{(\alpha)}(z)}{\Gamma(n+\alpha+1)}. \tag{10.1.4}$$

In particular, we have

$$\mathrm{e}^t\mathcal{J}_0\{2(zt)^{1/2}\} = \sum_{n=0}^{\infty}\frac{t^n L_n(z)}{n!}. \tag{10.1.5}$$

10.2 Rodrigues's formula

Using Leibniz's theorem for the nth derivative of a product, we have

$$\begin{aligned} \frac{\mathrm{d}^n}{\mathrm{d}z^n}(\mathrm{e}^{-z}z^{n+\alpha}) &= \sum_{r=0}^{n}\binom{n}{r}\frac{\mathrm{d}^r}{\mathrm{d}z^r}(\mathrm{e}^{-z})\frac{\mathrm{d}^{n-r}}{\mathrm{d}z^{n-r}}(z^{n+\alpha}) \\ &= \sum_{r=0}^{n}\frac{(-1)^r n!\,\mathrm{e}^{-z}\,\Gamma(n+\alpha+1)z^{\alpha+r}}{r!(n-r)!\,\Gamma(r+\alpha+1)}, \end{aligned}$$

and so it follows that

$$L_n^{(\alpha)}(z) = \frac{\mathrm{e}^z z^{-\alpha}}{n!}\frac{\mathrm{d}^n}{\mathrm{d}z^n}(\mathrm{e}^{-z}z^{n+\alpha}). \tag{10.2.1}$$

Hence, by Cauchy's formula we have

$$L_n^{(\alpha)}(z) = \frac{\mathrm{e}^z z^{-\alpha}}{2\pi\mathrm{i}}\int_C \frac{\mathrm{e}^{-\xi}\xi^{n+\alpha}}{(\xi - z)^{n+1}}\,\mathrm{d}\xi,$$

where C is a closed contour in the positive direction enclosing $\xi = z$ but excluding $\xi = 0$.

Another contour-integral expression for the Laguerre polynomial follows from the generating function. We have

$$L_n^{(\alpha)}(z) = \frac{n!}{2\pi i} \int_C \frac{(1-\xi)^{-\alpha-1}}{\xi^{n+1}} \exp\left(-\frac{z\xi}{1-\xi}\right) d\xi,$$

where C is a closed contour in the positive direction enclosing the origin.

10.3 Orthogonality relations

For real x we may write

$$L_n^{(\alpha)}(x) = e^x x^{-\alpha} \eta^{(n)}(x)/n!,$$

where $\eta^{(n)}(z)$ denotes the nth derivative of $\eta(x) \equiv e^{-x} x^{n+\alpha}$. We define

$$I_{m,n} \equiv \int_0^\infty x^\alpha\, e^{-x} L_m^{(\alpha)}(x) L_n^{(\alpha)}(x)\, dx$$
$$= \frac{1}{n!} \int_0^\infty L_m^{(\alpha)}(x) \eta^{(n)}\, dx.$$

There is no loss in generality in assuming that $m \leqslant n$. Successive integration by parts yields

$$I_{m,n} = \frac{(-1)^k}{n!} \int_0^\infty \frac{d^k}{dx^k}\{L_m^{(\alpha)}(x)\} \eta^{(n-k)}\, dx$$

provided that $\mathscr{R}(\alpha) > -1$, since $\eta^{(n-k)}$ vanishes at the limits 0 and ∞. For $k = m + 1$, the $(m + 1)$th derivative of $L_m^{(\alpha)}(x)$ vanishes, since $L_m^{(\alpha)}(x)$ is a polynomial of degree m. Accordingly $I_{m,n} = 0$ if $m < n$. For $m = n$, we have

$$I_{n,n} = \frac{(-1)^n}{n!} \int_0^\infty \frac{d^n}{dx^n}\{L_n^{(\alpha)}(x)\} \eta\, dx.$$

But

$$\frac{d^n}{dx^n}\{L_n^{(\alpha)}(x)\} = (-1)^n$$

and so

$$I_{n,n} = \frac{1}{n!} \int_0^\infty e^{-x} x^{n+\alpha}\, dx = \Gamma(1 + \alpha + n)/n!.$$

Hence we have the orthogonality and normalizing relations

$$\int_0^\infty x^\alpha e^{-x} L_m^{(\alpha)}(x) L_n^{(\alpha)}(x)\, dx = \Gamma(1 + \alpha + n)\delta_{mn}/n!, \qquad (10.3.1)$$

where $\delta_{mn} = 1$ when $m = n$, but otherwise vanishes.

10.4 The Laguerre differential equation

From $\Phi = e^{-z}z^{n+\alpha}$ we obtain

$$z\Phi' - (n + \alpha - z)\Phi = 0.$$

The result of differentiating $n + 1$ times is

$$z\Phi^{(n+2)} + (z + 1 - \alpha)\Phi^{(n+1)} + (n + 1)\Phi^{(n)} = 0,$$

and so $\Psi = d^n(e^{-z}z^{n+\alpha})/dz^n$ satisfies the equation

$$z\Psi'' + (z + 1 - \alpha)\Psi' + (n + 1)\Psi = 0.$$

The substitution $\Psi = n!\, e^{-z}z^{\alpha}w$ now yields

$$zw'' + (\alpha + 1 - z)w' + nw = 0, \qquad (10.4.1)$$

and so the Laguerre polynomial $L_n^{(\alpha)}(z)$ satisfies this differential equation.

The equation

$$zw'' + (\alpha + 1 - z)w' + \nu w = 0, \qquad (10.4.2)$$

where ν need no longer be integral, is known as Laguerre's differential equation.

The reader's attention is drawn to the fact that a different nomenclature is often employed in the literature of quantum mechanics. There $L_n^{(\alpha)}(z)$ stands for $d^{\alpha}L_n(z)/dz^{\alpha}$ and is usually called the associated Laguerre polynomial. This polynomial satisfies the differential equation

$$zw'' + (\alpha + 1 - z)w' + (n - \alpha)w = 0.$$

The solutions of equation (10.4.2) for non-integral α were discussed in illustration 3 of Chapter 1, and two independent solutions (1.4.4) and (1.4.5) were obtained. We now show that the only solution of (10.4.2) of interest in quantum mechanics is the Laguerre polynomial.

The Laguerre differential equation arises in quantum mechanics when Schrödinger's equation for the hydrogen atom is solved by the method of separation of variables in spherical polar coordinates r, θ and ϕ. With the usual physical meaning attached to the symbols m, e, Z, h and $E\,(< 0)$, the differential equation in r takes the form

$$R'' + \frac{2}{r}R' + \left(A + \frac{2B}{r} + \frac{C}{r^2}\right)R = 0, \qquad (10.4.3)$$

where primes now denote derivatives with respect to r, $A = 8\pi^2 mE/h^2$, $B = 4\pi^2 mZe^2/h^2$ and $C = -l(l + 1)$, l being a positive integer.

The transformation $R = e^{-r/2}r^l w(r)$ now yields the Laguerre equation

$$rw'' + (2l + 2 - r)w' + (\rho - l - 1)w = 0, \qquad (10.4.4)$$

where $\rho^2 = -2\pi^2 mZ^2e^4/Eh^2 > 0$.

The solution $R(r)$ of (10.4.3) in accordance with the physical conditions implied must satisfy the requirements

(i) $R(r)$ is bounded as $r \to 0$,

(ii) $R(r) \to 0$ as $r \to \infty$.

Equation (10.4.4) is the confluent hypergeometric equation (7.1.6) with $c = 2l + 2$ and $a = l + 1 - \rho$ and so has the independent solutions $r^{-l-1}\, e^{r/2} W_{\rho,l+1/2}(r)$ and $r^{-l-1}\, e^{r/2} W_{-\rho,l+1/2}(-r)$. It follows that (10.4.3) has the independent solutions $r^{-1}W_{\rho,l+1/2}(r)$ and $r^{-1}W_{-\rho,l+1/2}(-r)$. Then by (7.7.2) these solutions are of the order $e^{-r/2}O(r^{\rho-1})$ and $e^{r/2}O(r^{-\rho-1})$ at infinity. Hence, by condition (ii), the only acceptable solution is $R(r) = r^{-1}W_{\rho,l+1/2}(r)$.

Since c is a positive integer greater than unity, we may take the solution of (10.4.3) in the form

$$e^{-r/2} r^l U(a; c; r)/\Gamma(a),$$

where $U(a; c; r)$ is given by (7.4.6) and $\Gamma(a)$ is introduced in the denominator to remove the singularities which would otherwise occur when a was a negative integer. This gives

$$\begin{aligned} e^{r/2}R(r) = & \sum_{s=0}^{c-2} \frac{\Gamma(1 + a - c + s)\, \Gamma(c - s - 1)}{\Gamma(1 + a - c)\, \Gamma(a) s!} r^{s-c/2} \\ & + (-1)^c \frac{\log r\, {}_1F_1(a; c; r) r^{c/2-1}}{\Gamma(c)\, \Gamma(1 + a - c)} \\ & + (-1)^c \frac{1}{\Gamma(1 + a - c)\, \Gamma(a)} \sum_{s=0}^{\infty} \frac{\Gamma(a + s)}{\Gamma(c + s)} \\ & \times \{\psi(a + s) - \psi(c + s) - \psi(1 + s)\} \frac{r^{s+c/2-1}}{s!}, \end{aligned}$$

where $\psi(z) = \Gamma'(z)/\Gamma(z)$.

The first term in this summation is $\Gamma(c - 1)r^{-c/2}/\Gamma(a)$, and so condition (i) is satisfied only if a is chosen so that $1/\Gamma(a)$ vanishes. Thus a is a negative integer or zero, and the entire first summation on the right-hand side vanishes.

In this case $1/\Gamma(1 + a - c)$ also vanishes, and so the terms containing $\log r$ also disappear. Accordingly ρ is an integer and $\rho \geqslant l + 1$. Further, the required solution for R is given by

$$\begin{aligned} e^{r/2}R(r) & = \lim_{a+N\to 0} \frac{(-1)^c}{\Gamma(1 + a - c)\, \Gamma(a)} \sum_{s=0}^{\infty} \frac{\Gamma(a + s)}{\Gamma(c + s)} \frac{\psi(a + s)}{s!} r^{s+c/2-1} \\ & = \lim_{a+N\to 0} \frac{(-1)^c}{\Gamma(1 + a - c)\, \Gamma(a)} \sum_{s=0}^{N} \frac{\Gamma(a + s)}{\Gamma(c + s)} \frac{\psi(a + s)}{s!} r^{s+c/2-1} \end{aligned}$$

where N is a positive integer.

We have

$$\lim_{a+N\to 0} \frac{\Gamma(a+s)}{\Gamma(a)} = (s-N-1)(s-N-2)\ldots(-N)$$
$$= (-1)^s N!/(N-s)!$$

and

$$\lim_{a+N\to 0} \frac{\psi(a+s)}{\Gamma(1+a-c)}$$
$$= \lim_{a+N\to 0} \frac{\Gamma(c+N)}{\Gamma(1+a-c)} \frac{\psi(a+s)}{\Gamma(c-a)}$$
$$= \Gamma(c+N) \lim_{a+N\to 0} \frac{\sin \pi(c-a)}{\pi} \left\{-\gamma - \frac{1}{a+s} + \sum_{r=1}^{\infty} \left(\frac{1}{r} - \frac{1}{r+a+s}\right)\right\}$$
$$- (-1)^{c+N}\,\Gamma(c+N) \lim_{a+N\to 0} \frac{\sin \pi(a+N)}{\pi} \sum_{r=1}^{\infty} \left(\frac{1}{r} - \frac{1}{r+a+s}\right)$$

But one and only one of the expressions $r + a + s$ is equal to $a + N$, and so the limit is -1. Hence the desired solution is given by

$$-r^{-l} e^{r/2} R(r) = (-1)^N\,\Gamma(c+N) \sum_{s=0}^{N} \frac{(1-)^s N!\, r^s}{(N-s)!\,\Gamma(2l+2+s)s!.}$$

We readily from (10.1.2) identify the right-hand side as a multiple of the Laguerre polynomial $L_{\rho-l-1}{}^{2l+1}(r)$.

10.5 Recurrence relations

From (10.1.1) we have the relations

$$(1-t)\frac{\partial G}{\partial z} + tG = 0,$$

and

$$\{z - 1 - \alpha + (1+\alpha)t\} \frac{\partial G}{\partial z} = t(1-t)\frac{\partial G}{\partial t}.$$

On equating coefficients of t^n we obtain

$$\mathrm{D}L_n^{(\alpha)}(z) - \mathrm{D}L_{n-1}^{(\alpha)}(z) + L_{n-1}^{(\alpha)}(z) = 0 \qquad (10.5.1)$$

and

$$(z - 1 - \alpha)\mathrm{D}L_n^{(\alpha)}(z) + (1+\alpha)\mathrm{D}L_{n-1}^{(\alpha)}(z)$$
$$= nL_n^{(\alpha)}(z) - (n-1)L_{n-1}^{(\alpha)}(z), \qquad (10.5.2)$$

where $\mathrm{D} = \mathrm{d}/\mathrm{d}z$. The addition of $(1 + \alpha)$ times (10.5.1) to (10.5.2) yields

$$z\mathrm{D}L_n^{(\alpha)}(z) = nL_n^{(\alpha)}(z) - (\alpha + n)L_{n-1}^{(\alpha)}(z). \qquad (10.5.3)$$

Replacing n by $n-1$, we have

$$zDL_{n-1}^{(\alpha)}(z) = (n-1)L_{n-1}^{(\alpha)}(z) - (\alpha+n-1)L_{n-2}^{(\alpha)}(z). \quad (10.5.4)$$

Elimination of $DL_n^{(\alpha)}(z)$ and $DL_{n-1}^{(\alpha)}(z)$ from (10.5.1), (10.5.3) and (10.5.4) now gives

$$nL_n^{(\alpha)}(z) = (2n-1+\alpha-z)L_{n-1}^{(\alpha)}(z) - (n-1+\alpha)L_{n-2}^{(\alpha)}(z), \quad (10.5.5)$$

which is a recurrence relation between three successive Laguerre polynomials with the same α.

Again from (10.1.1) we have

$$\sum_{n=0}^{\infty} L_n^{(\alpha)}(z)t^n = (1-t)\sum_{n=0}^{\infty} L_n^{(\alpha+1)}(z)t^n$$

and so

$$L_n^{(\alpha)}(z) = L_n^{(\alpha+1)}(z) - L_{n-1}^{(\alpha+1)}(z). \quad (10.5.6)$$

It follows that

$$\sum_{r=1}^{n} L_r^{(\alpha)}(z) = L_n^{(\alpha+1)}(z) - 1$$

and this relation can be rewritten in the form

$$L_n^{(\alpha+1)}(z) = \sum_{r=0}^{n} L_r^{(\alpha)}(z). \quad (10.5.7)$$

Further, from (10.1.1) we have

$$\frac{\partial G}{\partial z} = -t(1-t)^{-1}G = -\sum_{r=1}^{\infty} t^r G$$

and so it follows that

$$DL_n^{(\alpha)}(z) = -\sum_{s=0}^{n-1} L_s^{(\alpha)}(z). \quad (10.5.8)$$

In combination with (5.7) we have

$$DL_n^{(\alpha)}(z) = -L_{n-1}^{(\alpha+1)}(z). \quad (10.5.9)$$

This result could have been deduced directly from (10.1.2).

10.6 Addition theorems

We have from the generating function that

$$\begin{aligned}\sum_{n=0}^{\infty} L_n^{(\alpha_1+\alpha_2+1)}(z_1+z_2)t^n &= (1-t)^{-2-\alpha_1-\alpha_2}\exp\{-(z_1+z_2)t/(1-t)\} \\ &= (1-t)^{-1-\alpha_1}\exp\{-z_1t/(1-t)\} \\ &\quad \times (1-t)^{-1-\alpha_2}\exp\{-z_2t/(1-t)\} \\ &= \sum_{r=0}^{\infty} L_r^{(\alpha_1)}(z_1)t^r \sum_{s=0}^{\infty} L_s^{(\alpha_2)}(z_2)t^s\end{aligned}$$

and so it follows that

$$L_n^{(\alpha_1+\alpha_2+1)}(z_1 + z_2) = \sum_{r+s=n} L_r^{(\alpha_1)}(z_1)L_s^{(\alpha_2)}(z_2). \tag{10.6.1}$$

Similarly, we obtain the generalized addition theorem

$$L_n^{(A)}\left(\sum_{i=1}^{k} z_i\right) = \sum_{\Sigma r_i = n} \left\{\prod_{i=1}^{k} L_{r_i}^{(\alpha_i)}(z_i)\right\}, \tag{10.6.2}$$

where $A = \sum_{i=1}^{k} \alpha_i + k - 1.$

Other interesting results can be obtained in this manner. For example,

$$\sum_{n=0}^{\infty} L_n^{(\alpha)}(z)t^n = (1-t)^{-(\alpha-\beta)} \sum_{r=0}^{\infty} L_r^{(\beta)}(z)t^r$$

and so

$$\Gamma(\alpha - \beta)L_n^{(\alpha)}(z) = \sum_{r=0}^{n} \frac{\Gamma(\alpha - \beta + r)}{r!} L_{n-r}^{(\beta)}(z). \tag{10.6.3}$$

10.7 Integral representation in terms of Bessel functions

Consider

$$\begin{aligned}
&\frac{1}{n!}\int_0^\infty e^{-t}t^{n+\alpha/2}J_\alpha\{2(tz)^{1/2}\}\,dt \\
&= \frac{1}{n!}\int_0^\infty e^{-t}t^{n+\alpha/2}\sum_{r=0}^{\infty}\frac{(-1)^r(tz)^{\alpha/2+r}}{r!\,\Gamma(r+\alpha+1)}\,dt \\
&= \frac{1}{n!}\sum_{r=0}^{\infty}\frac{(-1)^r z^{\alpha/2+r}}{r!\,\Gamma(r+\alpha+1)}\int_0^\infty e^{-t}t^{n+r+\alpha}\,dt \\
&= \frac{1}{n!}\sum_{r=0}^{\infty}\frac{(-1)^r z^{\alpha/2+r}\,\Gamma(n+r+\alpha+1)}{r!\,\Gamma(r+\alpha+1)} \\
&= \frac{z^{\alpha/2}\,\Gamma(n+\alpha+1)}{n!\,\Gamma(\alpha+1)}\,{}_1F_1(n+\alpha+1;\alpha+1;-z), \quad \text{by (7.1.3)} \\
&= \frac{z^{\alpha/2}\,\Gamma(n+\alpha+1)}{n!\,\Gamma(\alpha+1)}\,e^{-z}{}_1F_1(-n;\alpha+1;z), \quad \text{by (7.3.3)} \\
&= z^{\alpha/2}e^{-z}L_n^{(\alpha)}(z), \quad \text{by (10.1.3)}.
\end{aligned}$$

Hence we have established the integral representation

$$L_n^{(\alpha)}(z) = \frac{z^{-\alpha/2}\,e^z}{n!}\int_0^\infty e^{-t}t^{n+\alpha/2}J_\alpha\{2(tz)^{1/2}\}\,dt. \tag{10.7.1}$$

10.8 Expansions in series of Laguerre polynomials

It can be shown that the functions $x^{\alpha/2}\,\mathrm{e}^{-x/2}L_n^{(\alpha)}(x)$ form a complete orthogonal system for the range $(0, \infty)$ (see section 3.10), and so any function which is well behaved can be expanded in a series of Laguerre polynomials.

For the polynomial x^n it is easier to obtain the expansion without recourse to (10.3.1). From the equation which we used to derive (10.1.4) we note that

$$\left(\sum_{n=0}^{\infty}\frac{t^n}{n!}\right)\left(\sum_{r=0}^{\infty}\frac{(-1)^r(zt)^r}{r!\,\Gamma(r+\alpha+1)}\right)=\sum_{n=0}^{\infty}\frac{t^nL_n^{(\alpha)}(z)}{\Gamma(n+\alpha+1)}.$$

Accordingly,

$$\sum_{n=0}^{\infty}\frac{(-1)^n(zt)^n}{n!\,\Gamma(n+\alpha+1)}=\sum_{n=0}^{\infty}\frac{(-1)^nt^n}{n!}\sum_{r=0}^{\infty}\frac{t^rL_r^{(\alpha)}(z)}{\Gamma(r+\alpha+1)}$$

$$=\sum_{n=0}^{\infty}\sum_{r=0}^{n}\frac{(-1)^{n-r}t^nL_r^{(\alpha)}(z)}{(n-r)!\,\Gamma(r+\alpha+1)},$$

and so by comparison of the coefficients of t^n we have

$$z^n=\sum_{r=0}^{n}\frac{(-1)^rn!\,\Gamma(n+\alpha+1)\,L_r^{(\alpha)}(z)}{(n-r)!\,\Gamma(r+\alpha+1)}. \tag{10.8.1}$$

Exercises

1. Show that the Laguerre polynomials can be defined by the generating function

$$\mathrm{e}^{-tz}(1+t)^{\alpha}=\sum_{n=0}^{\infty}L_n^{(\alpha-n)}(z)t^n.$$

2. Establish the recurrence relation

$$(n+1)L_{n+1}^{(\alpha)}(z)=(n+1-z)L_n^{(\alpha)}(z)+(n+\alpha)L_n^{(\alpha-1)}(z).$$

3. Show that the Laguerre polynomials satisfy the differential relations:

(a) $$\mathrm{D}^m\{\mathrm{e}^{-x}x^{\alpha}L_n^{(\alpha)}(x)\}=\frac{(m+n)!}{n!}\,\mathrm{e}^{-x}x^{\alpha-m}L_{(m+n)}^{(\alpha-m)}(x);$$

(b) $$\mathrm{D}^m\{x^{\alpha}L_n^{(\alpha)}(x)\}=\frac{\Gamma(n+\alpha+1)}{\Gamma(n-m+\alpha)}x^{\alpha-m}L_n^{(\alpha-m)}(x).$$

4. Prove that

$$\lim_{n\to\infty}\left\{n^{-\alpha}L_n^{(\alpha)}\left(\frac{x}{n}\right)\right\}=x^{-\alpha/2}J_{\alpha}(2x^{1/2}).$$

[*Hint:* $\log\Gamma(z)=(z-\tfrac{1}{2})\log z-z+\tfrac{1}{2}\log(2\pi)+O(1/|z|)$ for large $|z|$.]

5. Show that the Laplace transform of $L_n^{(\alpha)}(t)$, where $\mathscr{R}(\alpha) > -1$, is

$$\Gamma(\alpha + n + 1)(s - 1)^n/(n!\, s^{\alpha+n+1}).$$

Hence deduce that the Laplace transform of $L_n(t)$ is $(1 - 1/s)^n/s$ and then use the convolution theorem to show that

$$\int_0^x L_m(y)L_n(x-y)\,\mathrm{d}y = \int_0^x L_{m+n}(y)\,\mathrm{d}y$$
$$= L_{m+n}(x) - L_{m+n+1}(x).$$

6. Establish the relation

$$L_n(xy) = \sum_{s=0}^{n} \binom{n}{s}(1-y)^{n-s}y^s L_s(x).$$

[*Hint:* use (10.1.5).]

7. Show that

$$\int_0^\infty \mathrm{e}^{-x}x^{s+\alpha}L_n^{(\alpha)}(x)\,\mathrm{d}x = 0 \quad \text{if } s < n, \text{ but}$$
$$= (-1)^n\binom{s}{n}\Gamma(s+\alpha+1) \quad \text{if } s > n.$$

CHAPTER 11

Hermite Polynomials

11.1 The generating function

The Hermite polynomials $H_n(z)$ are defined for non-negative integers n by the generating function

$$G(z, t) \equiv \exp(2zt - t^2) \equiv \sum_{n=0}^{\infty} H_n(z)t^n/n!. \qquad (11.1.1)$$

We have

$$\begin{aligned} G(z, t) &= \exp(2zt)\exp(-t^2) \\ &= \sum_{n=0}^{\infty} \frac{2^n z^n t^n}{n!} \sum_{r=0}^{\infty} (-1)^r \frac{t^{2r}}{r!} \\ &= \sum_{n=0}^{\infty} \sum_{r=0}^{[n/2]} (-1)^r (2z)^{n-2r} t^n / \{r!(n-2r)!\} \end{aligned}$$

and so

$$H_n(z) = \sum_{r=0}^{[n/2]} \frac{(-1)^r n!(2z)^{n-2r}}{r!(n-2r)!}. \qquad (11.1.2)$$

It follows that $H_n(z)$ is a polynomial of degree n in z which is even or odd according as n is even or odd respectively. That is,

$$H_n(-z) = (-1)^n H_n(z).$$

We verify that

$$\begin{aligned} &H_0(z) = 1, \quad H_1(z) = 2z, \quad H_2(z) = 4z^2 - 2, \\ &H_3(z) = 8z^3 - 12z, \quad H_4(z) = 16z^4 - 48z^2 + 12; \\ &H_{2n}(0) = (-1)^n(2n)!/n!, \quad H_{2n+1}(0) = 0. \end{aligned}$$

It is useful to display the Hermite polynomials as series in ascending powers of z. The results are expressed separately for $H_{2n}(z)$ and $H_{2n+1}(z)$ in the forms

$$H_{2n}(z) = \sum_{r=0}^{n} \frac{(-1)^{n+r}(2n)!(2z)^{2r}}{(n-r)!\,(2r)!} = (-1)^n(2n)!\,{}_1F_1(-n; \tfrac{1}{2}; z^2)/n!$$
$$= (-1)^n 2^{2n} n! L_n^{(-1/2)}(z^2) \qquad (11.1.3)$$

and

$$H_{2n+1}(z) = \sum_{r=0}^{n} \frac{(-1)^{n+r}(2n+1)!(2z)^{2r+1}}{(n-r)!\,(2r+1)!} = (-1)^n(2n+1)!2z$$
$$\times\, {}_1F_1(-n; \tfrac{3}{2}; z^2)/n!$$
$$= (-1)^n 2^{2n+1} n! z L_n^{(1/2)}(z^2). \qquad (11.1.4)$$

11.2 Rodrigues's formula

From the defining generating function (11.1.1) we have by Maclaurin's theorem that

$$H_n(z) = \left[\frac{d^n}{dt^n}\exp(2zt - t^2)\right]_{t=0}$$
$$= \exp(z^2)\left[\frac{d^n}{dt^n}\exp\{-(z-t)^2\}\right]_{t=0}$$
$$= (-1)^n \exp(z^2)\left[\frac{d^n}{dw^n}\exp(-w^2)\right]_{w=z}$$

and so

$$H_n(z) = (-1)^n \exp(z^2)\frac{d^n}{dz^n}\exp(-z^2). \qquad (11.2.1)$$

Hence by Cauchy's formula we have

$$H_n(z) = \frac{(-1)^n n! \exp(z^2)}{2\pi i}\int_C \frac{\exp(-\xi^2)\, d\xi}{(\xi - z)^{n+1}}$$

where C is a closed contour in the positive direction enclosing the point $\xi = z$.

Another contour-integral expression for the Hermite polynomial follows from the generating function. We have

$$H_n(z) = \frac{n!}{2\pi i}\int_C \xi^{-n-1}\exp(2z\xi - \xi^2)\, d\xi,$$

where C is a closed contour in the positive direction enclosing the origin. Choose C to be the unit circle on which $\xi = e^{i\theta}$ and we obtain

$$H_n(z) = \frac{n!}{2\pi}\int_0^{\pi}\exp(2z\cos\theta - \cos 2\theta)\cos(2z\sin\theta - \sin 2\theta - n\theta)\, d\theta.$$

11.3 Orthogonal properties

Consider for real x the integrals

$$I_{m,n} \equiv \int_{-\infty}^{\infty} \exp(-x^2) H_m(x) H_n(x)\, dx,$$

where there is no loss in generality in assuming that $m \leqslant n$. Using Rodrigues's formula (11.2.1), we have

$$I_{m,n} = (-1)^n \int_{-\infty}^{\infty} H_m(x) \frac{d^n}{dx^n} \exp(-x^2)\, dx.$$

Integration by parts yields

$$I_{m,n} = \left[(-1)^n H_m(x) \frac{d^{n-1}}{dx^{n-1}} \exp(-x^2)\right]_{-\infty}^{\infty} + (-1)^{n+1} \int_{-\infty}^{\infty} \frac{d}{dx} H_m(x) \frac{d^{n-1}}{dx^{n-1}} \exp(-x^2)\, dx.$$

The integrated portion is the product of $\exp(-x^2)$ and a polynomial in x, and so vanishes at both limits. After m integrations by parts we obtain

$$I_{m,n} = (-1)^{n+m} \int_{-\infty}^{\infty} \frac{d^m}{dx^m} H_m(x) \frac{d^{n-m}}{dx^{n-m}} \exp(-x^2)\, dx.$$

In the case $m < n$, one more integration by parts produces $d^{m+1} H_m(x)/dx^{m+1}$ in the integrand. This expression is zero since $H_m(x)$ is a polynomial of degree m. Accordingly $I_{m,n} = 0$ for $m \neq n$.

In the case $m = n$ we have

$$I_{m,n} = \int_{-\infty}^{\infty} \frac{d^n}{dx^n} H_n(x) \exp(-x^2)\, dx.$$

But $H_n(x)$ is a polynomial of degree n in which, from (11.1.2), the coefficient of x^n is 2^n. Hence

$$I_{n,n} = 2^n n! \int_{-\infty}^{\infty} \exp(-x^2)\, dx = 2^n n! \sqrt{\pi},$$

and so we obtain the orthogonality relations

$$I_{m,n} = 2^n n! \sqrt{\pi} \delta_{mn}. \tag{11.3.1}$$

11.4 The Hermite differential equation

The function $\eta = \exp(-z^2)$ satisfies the differential equation

$$\frac{d\eta}{dz} + 2z\eta = 0.$$

Differentiating $(n + 1)$ times, we obtain

$$\frac{d^2\xi}{dz^2} + 2z \frac{d\xi}{dz} + 2(n + 1)\xi = 0 \tag{11.4.1}$$

where $\xi = \mathrm{d}^n \exp(-z^2)/\,\mathrm{d}z^n$. The substitution $\xi = (-1)^n w \exp(-z^2)$ now yields the differential equation

$$\frac{\mathrm{d}^2 w}{\mathrm{d}z^2} - 2z\frac{\mathrm{d}w}{\mathrm{d}z} + 2nw = 0,$$

and so $H_n(z)$ satisfies this differential equation. The equation

$$w'' - 2zw' + 2\nu w = 0, \tag{11.4.2}$$

where ν need no longer be integral, is known as Hermite's differential equation.

Except for the point at infinity, all points are ordinary points of the Hermite equation. This equation for non-integer n (ν, say) was investigated in illustration 2 of Chapter 1, and the two solutions

$$w_1(z) = \Gamma(\tfrac{1}{2}\nu + 1, \sum_0^\infty (-1)^s (2z)^{2s}/\{(2s)!\,\Gamma(\tfrac{1}{2}\nu - s + 1)\},$$

$$w_2(z) = \tfrac{1}{2}\Gamma(\tfrac{1}{2}\nu + \tfrac{1}{2}) \sum_0^\infty (-1)^s (2z)^{2s+1}/\{(2s+1)!\,\Gamma(\tfrac{1}{2}\nu - s + \tfrac{1}{2})\}$$

were obtained. These functions are sometimes called the Hermite functions of the second kind and are clearly entire functions. We note that $w_1(z)$ is a polynomial of degree n when ν is an even integer n, and that $w_2(z)$ is a polynomial of degree n when ν is an odd integer n. These polynomials multiplied respectively by appropriate constants are the Hermite polynomials.

At this stage it is instructive to show why the solutions corresponding to the quantum-mechanical discussion of the linear harmonic oscillator are Hermite polynomials of odd degree. The Schrödinger wave function ψ in this case satisfies the differential equation

$$\frac{\mathrm{d}^2\psi}{\mathrm{d}x^2} + (\lambda - \alpha^2 x^2)\psi = 0, \tag{11.4.3}$$

where x is the distance displaced, $\lambda = 8\pi^2 mE/h^2$, $\alpha = 2\pi m\omega_0/h > 0$, m, E, ω_0 and h having the usual physical significance.

The physical conditions impose the requirements that

(i) $\psi \to 0$ as $|x| \to \infty$,

(ii) $\psi \to 0$ as $x \to 0$.

The substitutions $\sqrt{\alpha}x = \xi$, $\psi = \exp(-\tfrac{1}{2}\xi^2)y$ yield Hermite's equation

$$\frac{\mathrm{d}^2 y}{\mathrm{d}\xi^2} - 2\xi\frac{\mathrm{d}y}{\mathrm{d}\xi} + 2\nu\xi = 0,$$

where $2\nu = \lambda/\alpha - 1$.

A further substitution $\xi^2 = u$ now produces the confluent hypergeometric equation

$$u\frac{d^2y}{du^2} + (c-u)\frac{dy}{du} - ay = 0,$$

where $c = \frac{1}{2}$ and $a = -\nu/2$.

This equation has the two independent solutions $u^{-1/4}e^{u/2}W_{k,m}(u)$ and $u^{-1/4}\,e^{u/2}W_{-k,m}(-u)$, where $k = \frac{1}{4} + \nu/2$ and $m = -\frac{1}{4}$. That is, the two solutions of (11.4.3) are $x^{-1/2}W_{k,m}(\alpha x^2)$ and $x^{-1/2}W_{-k,m}(-\alpha x^2)$. These solutions are asymptotic to $\exp(-\frac{1}{2}\alpha x^2)$ and $\exp(\frac{1}{2}\alpha x^2)$ respectively, and so the second solution must be discarded in virtue of condition (i). Accordingly from (7.7.2) the required solution is

$$\frac{1}{\Gamma(-\nu/2)}U(-\tfrac{1}{2}\nu;\tfrac{1}{2};\alpha x^2).$$

The function $U(a;c;z)$ is not, in general, single-valued. We see from (7.5.3) that any path from $+\infty$ to $-\infty$ involves a change in the argument of αx^2 and so introduces a term ${}_1F_1(a;c;\alpha x^2)$ which has been shown in section 7.8 to be exponentially large at infinity. To avoid this we must choose a and c so that $1/\Gamma(1+a-c) = 0$. That is, $1/\Gamma(-\frac{1}{2}-\nu/2) = 0$, and so $\nu = 2n+1$, where n is a positive integer.

The solution now takes the form

$$\frac{\Gamma(\frac{1}{2})}{\Gamma(-n-\frac{1}{2})}\alpha^{1/2}x\,{}_1F_1(-n;\tfrac{3}{2};\alpha x^2)$$

which is proportional, by (11.1.4), to the Hermite polynomial $H_{2n+1}(\alpha x^2)$.

Note carefully that condition (ii) is now satisfied automatically.

Further, equation (11.4.3) is closely related to Weber's differential equation

$$w'' + (\nu + \tfrac{1}{2} - \tfrac{1}{4}z^2)w = 0, \tag{11.4.4}$$

whose solutions are called parabolic cylinder functions. It is usual to choose the two solutions as $D_n(z)$ and $D_{-n-1}(z\,e^{\pi i/2})$, where

$$D_n(z) = 2^{\frac{1}{2}\nu+\frac{1}{4}}z^{-\frac{1}{2}}W_{\frac{1}{2}\nu+\frac{1}{4},-\frac{1}{4}}(\tfrac{1}{2}z^2).$$

The properties of the parabolic cylinder functions can be obtained from those of the confluent hypergeometric functions. In particular, we note that

$$D_n(z) = 2^{-n/2}\exp(-\tfrac{1}{4}z^2)H_n(2^{-1/2}z).$$

11.5 Recurrence relations

The differential equation (11.4.1) can be written in the form

$$\frac{d^{n+2}}{dz^{n+2}}\exp(-z^2) + 2z\frac{d^{n+1}}{dz^{n+1}}\exp(-z^2) + 2(n+1)\frac{d^n}{dz^n}\exp(-z^2) = 0.$$

Multiplication by $(-1)^{n+2}\exp(z^2)$ and use of Rodrigues's formula (11.2.1) immediately yields the recurrence relation

$$H_{n+2}(z) - 2zH_{n+1}(z) + 2(n+1)H_n(z) = 0. \tag{11.5.1}$$

From the generating function (11.1.1) we have

$$\frac{\partial G}{\partial z} = 2t\exp(2zt - t^2) = 2\sum_{n=0}^{\infty}\frac{H_n(z)t^{n+1}}{n!} = \sum_{n=0}^{\infty}\frac{\mathrm{d}H_n(z)}{\mathrm{d}z}\frac{t^n}{n!}$$

and so we derive

$$\frac{\mathrm{d}H_n(z)}{\mathrm{d}z} = 2nH_{n-1}(z). \tag{11.5.2}$$

In conjunction with (11.3.1) we can use these recurrence relations to evaluate integrals of the type

$$\int_{-\infty}^{\infty} x\exp(-x^2)H_m(x)H_n(x)\,\mathrm{d}x \quad\text{and}\quad \int_{-\infty}^{\infty}\exp(-x^2)H_m(x)\frac{\mathrm{d}H_n(x)}{\mathrm{d}x}\,\mathrm{d}x.$$

11.6 Addition theorems

From the generating function (11.1.1) we have

$$\begin{aligned}\sum_{n=0}^{\infty} H_n\left\{\frac{a_1z_1 + a_2z_2}{(a_1^2+a_2^2)^{1/2}}\right\}\frac{t^n}{n!} &= \exp\left[2\left\{\frac{a_1z_1+a_2z_2}{(a_1^2+a_2^2)^{1/2}}\right\}t - t^2\right]\\ &= \exp\left\{\frac{2a_1z_1t}{(a_1^2+a_2^2)^{1/2}} - \frac{a_1^2t^2}{a_1^2+a_2^2}\right\}\\ &\qquad\times\exp\left\{\frac{2a_2z_2t}{(a_1^2+a_2^2)^{1/2}} - \frac{a_2^2t^2}{a_1^2+a_2^2}\right\}\\ &= \sum_{r=0}^{\infty} H_r(z_1)\frac{a_1^rt^r}{(a_1^2+a_2^2)^{r/2}r!}\\ &\qquad\times\sum_{s=0}^{\infty} H_s(z_2)\frac{a_2^st^s}{(a_1^2+a_2^2)^{s/2}s!},\end{aligned}$$

and so we obtain the addition theorem

$$\frac{(a_1^2+a_2^2)^{n/2}}{n!}H_n\left\{\frac{a_1z_1+a_2z_2}{(a_1^2+a_2^2)^{1/2}}\right\} = \sum_{r+s=n}\frac{a_1^ra_2^s}{r!\,s!}H_r(z_1)H_s(z_2). \tag{11.6.1}$$

Similarly we can obtain the generalized addition theorem

$$\left(\sum_{r=1}^{m} a_r^2\right)^{n/2} H_n\left\{\frac{\sum_{r=1}^{m} a_rz_r}{\left(\sum_{r=1}^{m} a_r^2\right)^{1/2}}\right\} = n!\sum_{\Sigma p_r=n}\left\{\prod_{r=1}^{m}\frac{a_r^{p_r}}{(p_r)!}H_{p_r}(z_r)\right\}. \tag{11.6.2}$$

Putting $a_1 = a_2 = 1$ in (11.6.1), we have

$$2^{n/2}H_n\{(z_1+z_2)/\sqrt{2}\} = \sum_{r=0}^{n} \binom{n}{r} H_r(z_1)H_{n-r}(z_2). \tag{11.6.3}$$

Further, with $z_1 = z_2 = x$, we have

$$2^{n/2}H_n(\sqrt{2}x) = \sum_{r=0}^{n} \binom{n}{r} H_r(x)H_{n-r}(x). \tag{11.6.4}$$

Applications of (11.3.1) now yield

$$\int_{-\infty}^{\infty} \exp(-x^2)H_{2n+1}(\sqrt{2}x)\,dx = 0$$

and

$$\int_{-\infty}^{\infty} \exp(-x^2)H_{2n}(\sqrt{2}x)\,dx = (2n)!\,\sqrt{\pi}/n!.$$

The first result is trivial and the second may be rewritten

$$\int_{-\infty}^{\infty} \exp(-\tfrac{1}{2}x^2)H_{2n}(x)\,dx = (2n)!\,\sqrt{(\pi/2)}/n!.$$

11.7 Expansion in series of Hermite polynomials

It can be shown that the functions

$$\exp(-\tfrac{1}{2}x^2)H_n(x)/\{2^n n!\,\sqrt{\pi}\}^{1/2}$$

form a complete orthonormal system for the range $(-\infty, \infty)$, and so any well-behaved function can be expanded in a series of Hermite polynomials.

For the polynomial x^n it is easier to obtain the expansion without recourse to the orthonormal properties. We have

$$\sum_{n=0}^{\infty} \frac{(2x)^n t^n}{n!} = \exp(2xt) = \exp(2xt - t^2)\exp t^2$$
$$= \sum_{n=0}^{\infty} \frac{H_n(x)t^n}{n!} \sum_{r=0}^{\infty} \frac{t^{2r}}{r!}$$

and so we obtain

$$x^n = \sum_{r=0}^{[n/2]} \frac{n!\,H_{n-2r}(x)}{2^n r!\,(n-2r)!}. \tag{11.7.1}$$

As a corollary we deduce that

$$\int_{-\infty}^{\infty} x^n \exp(-x^2)H_m(x)\,dx = 0, \qquad n = 0, 1, \ldots (m-1)$$

and

$$\int_{-\infty}^{\infty} x^n \exp(-x^2)H_{n-2k}(x)\,dx = 2^{-2k}n!\,\sqrt{\pi}/k!.$$

EXERCISES

1. Show that

$$n!\sqrt{\pi}P_n(x) = \int_{-\infty}^{\infty} \exp(-t^2)t^n H_n(tx)\, dt.$$

2. Show that

$$\int_0^{\infty} \{D_n(x)\}^2\, dx = (2\pi)^{1/2} n!.$$

3. Establish the following:

(a) $2(n+1)\int_0^x H_n(t)\, dt = H_{n+1}(x) - H_{n+1}(0)$;

(b) $\int_0^x \exp(-t^2)H_n(t)\, dt = H_{n-1}(0) - \exp(-x^2)H_{n-1}(x)$.

[*Hint:* Establish $xH_n'(x) = nH_{n-1}'(x) + nH_n(x)$ and use (11.5.2).]

4. Prove that

$$k!\int_{-\infty}^{\infty} \exp(-x^2)x^n H_{n-2k}(x)\, dx = 2^{-2k} n!\,\sqrt{\pi} \qquad \text{for } k = 0, 1, \ldots [n/2].$$

5. Show that

$$\exp(2zt - t^2)H_m(z-t) = \sum_{n=0}^{\infty} H_{n+m}(z)t^n/n!$$

and deduce that

$$\sum_{n=0}^{\infty} \frac{H_n(z)H_n(w)t^n}{n!} = (1-4t^2)^{-1/2} \exp\left\{\frac{4(wzt - w^2t^2 - z^2t^2)}{1-4t^2}\right\}.$$

6. Show that

$$\sum_{k=0}^{n} \frac{H_k(z)H_k(w)}{2^k k!} = \frac{H_{n+1}(z)H_n(w) - H_n(z)H_{n+1}(w)}{2^{n+1}n!\,(w-z)}.$$

7. Show that

$$L_n^{(\alpha)}(x) = \frac{\Gamma(n+\alpha+1)x^{-\alpha}}{\Gamma(n+\beta+1)\,\Gamma(\alpha-\beta)} \int_0^x (x-u)^{\alpha-\beta-1}u^{\beta}L_n^{(\beta)}(u)\, du$$

when $\alpha > \beta$.

By letting $\beta = -\frac{1}{2}$ and $u = x\cos^2\theta$, deduce that

$$L_n^{(\alpha)}(x) = \frac{(-1)^n\,\Gamma(n+\alpha+1)}{\sqrt{\pi}(2n)!\,\Gamma(\alpha+\frac{1}{2})} \int_0^{\pi} H_{2n}(\sqrt{x}\cos\theta)\sin^{2\alpha}\theta\, d\theta.$$

APPENDIX 1

The Laplace and Helmholtz Equations

The purpose of this appendix is to indicate the types of differential equation which arise when Helmholtz's partial differential equation

$$\frac{\partial^2\psi}{\partial x^2}+\frac{\partial^2\psi}{\partial y^2}+\frac{\partial^2\psi}{\partial z^2}+k^2\psi=0$$

is solved in various orthogonal coordinate systems by the method of separation of variables.

Let u_1, u_2, u_3 denote the coordinates for which

$$\mathrm{d}s^2=h_1{}^2\,\mathrm{d}u_1{}^2+h_2{}^2\,\mathrm{d}u_2{}^2+h_3{}^2\,\mathrm{d}u_3{}^2.$$

It can be shown [14] that the Helmholtz equation transforms into

$$\frac{\partial}{\partial u_1}\left(\frac{h_2h_3}{h_1}\frac{\partial\psi}{\partial u_1}\right)+\frac{\partial}{\partial u_2}\left(\frac{h_3h_1}{h_2}\frac{\partial\psi}{\partial u_2}\right)+\frac{\partial}{\partial u_3}\left(\frac{h_1h_2}{h_3}\frac{\partial\psi}{\partial u_3}\right)+k^2h_1h_2h_3\psi=0.$$

A solution of the form $\psi=U_1(u_1)U_2(u_2)U_3(u_3)$ is assumed. Substitution in the differential equation followed by division by ψ yields

$$\frac{1}{U_1}\frac{\mathrm{d}}{\mathrm{d}u_1}\left(\frac{h_2h_3}{h_3}\frac{\mathrm{d}U_1}{\mathrm{d}u_1}\right)+\frac{1}{U_2}\frac{\mathrm{d}}{\mathrm{d}u_2}\left(\frac{h_3h_1}{h_2}\frac{\mathrm{d}U_2}{\mathrm{d}u_2}\right)+\frac{1}{U_3}\frac{\mathrm{d}}{\mathrm{d}u_3}\left(\frac{h_1h_2}{h_3}\frac{\mathrm{d}U_3}{\mathrm{d}u_3}\right)$$
$$+k^2h_1h_2h_3=0. \quad (\text{A1.1})$$

For certain coordinate systems it is now possible to separate (A1.1) into three ordinary differential equations for the U_i as functions of the corresponding u_i.

In order to avoid repetition, we shall omit all equations which yield

$$\frac{1}{U_i}\frac{\mathrm{d}^2U_i}{\mathrm{d}u_i{}^2}=c_i,$$

where c_i is a constant. Other constants introduced will be denoted by α or β.

In each of the following cases, the coordinate system will be defined, the values of h_i will be noted and the equation corresponding to (A1.1) will then be written. This will be followed by the ordinary differential equations satisfied by U_i (omitting those of the type mentioned in the previous paragraph) and any relevant comments.

(i) CIRCULAR CYLINDRICAL COORDINATES

$$x = u_1 \cos u_2, \qquad y = u_1 \sin u_2, \qquad z = u_3.$$
$$h_1 = 1, \qquad h_2 = u_1, \qquad h_3 = 1.$$
$$\frac{1}{U_1}\frac{\mathrm{d}}{\mathrm{d}u_1}\left(u_1\frac{\mathrm{d}U_1}{\mathrm{d}u_1}\right) + \frac{1}{U_2}\frac{\mathrm{d}}{\mathrm{d}u_2}\left(\frac{1}{u_1}\frac{\mathrm{d}U_2}{\mathrm{d}u_2}\right) + \frac{1}{U_3}\frac{\mathrm{d}}{\mathrm{d}u_3}\left(u_1\frac{\mathrm{d}U_3}{\mathrm{d}u_3}\right) + k^2u_1 = 0.$$
$$u_1^2\frac{\mathrm{d}^2U_1}{\mathrm{d}u_1^2} + u_1\frac{\mathrm{d}U_1}{\mathrm{d}u_1} + \{(c_3 + k^2)u_1^2 + c_2\}U_1 = 0,$$
which is the Bessel or the modified Bessel equation.

(ii) SPHERICAL POLAR COORDINATES

$$x = u_1 \sin u_2 \cos u_3, \qquad y = u_1 \sin u_2 \sin u_3, \qquad z = u_1 \cos u_2.$$
$$h_1 = 1, \qquad h_2 = u_1, \qquad h_3 = u_1 \sin u_2.$$
$$\frac{1}{U_1}\frac{\mathrm{d}}{\mathrm{d}u_1}\left(u_1^2 \sin u_2\frac{\mathrm{d}U_1}{\mathrm{d}u_1}\right) + \frac{1}{U_2}\frac{\mathrm{d}}{\mathrm{d}u_2}\left(\sin u_2\frac{\mathrm{d}U_2}{\mathrm{d}u_2}\right) + \frac{1}{U_3}\frac{\mathrm{d}}{\mathrm{d}u_3}\left(\frac{1}{\sin u_2}\frac{\mathrm{d}U_3}{\mathrm{d}u_3}\right)$$
$$+ k^2u_1^2 \sin u_2 = 0.$$
$$u_1^2\frac{\mathrm{d}^2U_1}{\mathrm{d}u_1^2} + 2u_1\frac{\mathrm{d}U_1}{\mathrm{d}u_1} + (k^2u_1^2 - \alpha)U_1 = 0,$$
which by the substitution $U_1 = u_1^{-1/2}V_1$ becomes the Bessel equation for V_1 as a function of ku_1,
$$\frac{\mathrm{d}^2U_2}{\mathrm{d}u_2^2} + \cot u_2\frac{\mathrm{d}U_2}{\mathrm{d}u_2} + \left(\frac{c_3}{\sin^2 u_2} + \alpha\right)U_2 = 0,$$
which is the associated Legendre equation for U_2 as a function of $\cos u_2$.

(iii) PARABOLIC CYLINDRICAL COORDINATES

$$x = \tfrac{1}{2}(u_1^2 - u_2^2), \quad y = u_1u_2, \qquad z = u_3.$$
$$h_1 = h_2 = (u_1^2 + u_2^2)^{1/2}, \qquad h_3 = 1.$$
$$\frac{1}{U_1}\frac{\mathrm{d}^2U_1}{\mathrm{d}u_1^2} + \frac{1}{U_2}\frac{\mathrm{d}^2U_2}{\mathrm{d}u_2^2} + \frac{u_1^2 + u_2^2}{U_3}\frac{\mathrm{d}^2U_3}{\mathrm{d}u_3^2} + k^2(u_1^2 + u_2^2) = 0.$$
$$\frac{\mathrm{d}^2U_1}{\mathrm{d}u_1^2} + \{(c_3 + k^2)u_1^2 + \alpha\}U_1 = 0,$$
$$\frac{\mathrm{d}^2U_2}{\mathrm{d}u_2^2} + \{(c_3 + k^2)u_2^2 - \alpha\}U_2 = 0,$$
both of which are Weber's equation for the parabolic cylinder function.

(iv) Parabolic coordinates

$$x = u_1 u_2 \cos u_3, \qquad y = u_1 u_2 \sin u_3, \qquad z = \tfrac{1}{2}(u_1^2 - u_2^2).$$

$$h_1 = h_2 = (u_1^2 + u_2^2)^{1/2}, \qquad h_3 = u_1 u_2.$$

$$\frac{1}{U_1}\frac{d}{du_1}\left(u_1 u_2 \frac{dU_1}{du_1}\right) + \frac{1}{U_2}\frac{d}{du_2}\left(u_1 u_2 \frac{dU_2}{du_2}\right) + \frac{1}{U_3}\frac{d}{du_3}\left(\frac{u_1^2 + u_2^2}{u_1 u_2}\frac{dU_3}{du_3}\right) + k^2 u_1 u_2 (u_1^2 + u_2^2) = 0,$$

$$u_1^2 \frac{d^2 U_1}{du_1^2} + u_1 \frac{dU_1}{du_1} + (k^2 u_1^4 + \alpha u_1^2 + c_3) U_1 = 0,$$

$$u_2^2 \frac{d^2 U_2}{du_2^2} + u_2 \frac{dU_2}{du_2} + (k^2 u_2^4 + \alpha u_2^2 + c_3) U_2 = 0.$$

The substitutions $u_i^2 = v_i$ and $U_i = v_i^{-1/2} V_i$ yields Whittaker's equations for V_i as functions of v_i.

(v) Elliptic cylindrical coordinates

$$x = a \cosh u_1 \cos u_2, \qquad y = a \sinh u_1 \sin u_2, \qquad z = u_3.$$

$$h_1 = h_2 = a\,(\sinh^2 u_1 + \sin^2 u_2)^{1/2}, \qquad h_3 = 1.$$

$$\frac{1}{U_1}\frac{d^2 U_1}{du_1^2} + \frac{1}{U_2}\frac{d^2 U_2}{du_2^2} + \frac{1}{U_3}\frac{d}{du_3}\left\{a^2(\sinh^2 u_1 + \sin^2 u_2)\frac{dU_3}{du_3}\right\} + k^2 a^2 (\sinh^2 u_1 + \sin^2 u_2) = 0.$$

$$\frac{d^2 U_1}{du_1^2} + \{\alpha + a^2(k^2 + c_3)\sinh^2 u_1\} U_1 = 0,$$

$$\frac{d^2 U_2}{du_2^2} + \{-\alpha + a^2(k^2 + c_3)\sin^2 u_2\} U_2 = 0,$$

both of which are Mathieu equations. The solutions of these equations are not discussed in the text, and the reader is referred to [1] and [18].

(vi) Prolate spheroidal coordinates

$$x = a \sinh u_1 \sin u_2 \cos u_3, \qquad y = a \sinh u_1 \sin u_2 \sin u_3, \qquad z = a \cosh u_1 \cos u_2.$$

$$h_1 = h_2 = a\,(\cosh^2 u_1 - \cos^2 u_2)^{1/2}, \qquad h_3 = a \sinh u_1 \sin u_2.$$

$$\frac{1}{U_1}\frac{d}{du_1}\left(\sinh u_1 \sin u_2 \frac{dU_1}{du_1}\right) + \frac{1}{U_2}\frac{d}{du_2}\left(\sinh u_1 \sin u_2 \frac{dU_2}{du_2}\right) + \frac{1}{U_3}\frac{d}{du_3}\left(\frac{\cosh^2 u_1 - \cos^2 u_2}{\sinh u_1 \sin u_2}\frac{dU_3}{du_3}\right) + k^2 a^2 (\cosh^2 u_1 - \cos^2 u_2)\sinh u_1 \sin u_2 = 0.$$

The substitutions $v_1 = \cosh u_1, v_2 = \cos u_2, v_3 = u_3$ transform this equation into

$$\frac{1}{U_1}\frac{\mathrm{d}}{\mathrm{d}v_1}\left\{(v_1^2 - 1)\frac{\mathrm{d}U_1}{\mathrm{d}v_1}\right\} + \frac{1}{U_2}\frac{\mathrm{d}}{\mathrm{d}v_2}\left\{(1 - v_2^2)\frac{\mathrm{d}U_2}{\mathrm{d}v_2}\right\}$$
$$+ \frac{1}{U_3}\left(\frac{1}{v_1^2 - 1} + \frac{1}{1 - v_2^2}\right)\frac{\mathrm{d}^2U_3}{\mathrm{d}v_3^2} + k^2a^2(v_1^2 - v_2^2) = 0.$$

$$\frac{\mathrm{d}}{\mathrm{d}v_1}\left\{(v_1^2 - 1)\frac{\mathrm{d}U_1}{\mathrm{d}v_1}\right\} + \left(\frac{c_3}{v_1^2 - 1} + k^2a^2v_1^2 + \alpha\right)U_1 = 0,$$

$$\frac{\mathrm{d}}{\mathrm{d}v_2}\left\{(1 - v_2^2)\frac{\mathrm{d}U_2}{\mathrm{d}v_2}\right\} + \left(\frac{c_3}{1 - v_2^2} - k^2a^2v_2^2 - \alpha\right)U_2 = 0,$$

both of which are Legendre's associated equation when $k = 0$. When $k \neq 0$ they are spheroidal wave equations [1].

(vii) OBLATE SPHEROIDAL COORDINATES

$$x = a\cosh u_1 \cos u_2 \cos u_3, \qquad y = a\cosh u_1 \cos u_2 \sin u_3,$$
$$z = a\sinh u_1 \sin u_2.$$

$$h_1 = h_2 = a(\sinh^2 u_1 + \sin^2 u_2)^{1/2}, \qquad h_3 = a\cosh u_1 \cos u_2.$$

$$\frac{1}{U_1}\frac{\mathrm{d}}{\mathrm{d}u_1}\left(\cosh u_1 \cos u_2 \frac{\mathrm{d}U_1}{\mathrm{d}u_1}\right) + \frac{1}{U_2}\frac{\mathrm{d}}{\mathrm{d}u_2}\left(\cosh u_1 \cos u_2 \frac{\mathrm{d}U_2}{\mathrm{d}u_2}\right)$$
$$+ \frac{1}{U_3}\frac{\mathrm{d}}{\mathrm{d}u_3}\left(\frac{\sinh^2 u_1 + \sin^2 u_2}{\cosh u_1 \cos u_2}\frac{\mathrm{d}U_3}{\mathrm{d}u_3}\right)$$
$$+ k^2a^2(\sinh^2 u_1 + \sin^2 u_2)\cosh u_1 \cos u_2 = 0.$$

The substitutions $v_1 = \sinh u_1$, $v_2 = \sin u_2$, $v_3 = u_3$ transform this equation into

$$\frac{1}{U_1}\frac{\mathrm{d}}{\mathrm{d}v_1}\left\{(1 + v_1^2)\frac{\mathrm{d}U_1}{\mathrm{d}v_1}\right\} + \frac{1}{U_2}\frac{\mathrm{d}}{\mathrm{d}v_2}\left\{(1 - v_2^2)\frac{\mathrm{d}U_2}{\mathrm{d}v_2}\right\}$$
$$+ \frac{1}{U_3}\left(\frac{1}{1 - v_2^2} - \frac{1}{1 + v_1^2}\right)\frac{\mathrm{d}^2U_3}{\mathrm{d}v_3^2} + ka^2(v_1^2 + v_2^2) = 0.$$

$$\frac{\mathrm{d}}{\mathrm{d}v_1}\left\{(1 + v_1^2)\frac{\mathrm{d}U_1}{\mathrm{d}v_1}\right\} + \left(\frac{-c_3}{1 + v_1^2} + k^2a^2v_1^2 + \alpha\right)U_1 = 0,$$

$$\frac{\mathrm{d}}{\mathrm{d}v_2}\left\{(1 - v_2^2)\frac{\mathrm{d}U_2}{\mathrm{d}v_2}\right\} + \left(\frac{c_3}{1 - v_2^2} + k^2a^2v_2^2 - \alpha\right)U_2 = 0,$$

the second of which is the associated Legendre equation when $k = 0$. When $k \neq 0$ they are spheroidal wave equations [1].

(viii) TOROIDAL COORDINATES

$$x = a\sinh u_1 \cos u_3/\xi, \qquad y = a\sinh u_1 \sin u_3/\xi, \qquad z = a\sin u_2/\xi,$$

where $\xi = \cosh u_1 - \cos u_2$,

$$h_1 = h_2 = a/\xi, \qquad h_3 = a \sinh u_1/\xi,$$

$$\frac{\partial}{\partial u_1}\left(\frac{\sinh u_1}{\xi}\frac{\partial \psi}{\partial u_1}\right) + \frac{\partial}{\partial u_2}\left(\frac{\sinh u_1}{\xi}\frac{\partial \psi}{\partial u_2}\right) + \frac{1}{\xi \sinh u_1}\frac{\partial^2 \psi}{\partial u_3{}^2} + k^2 a^2 \frac{\sinh u_1}{\xi^3}\psi = 0.$$

The substitution $\psi = \xi^{1/2} V$ yields

$$\sinh u_1\left(\frac{\partial^2 V}{\partial u_1{}^2} + \frac{\partial^2 V}{\partial u_2{}^2}\right) + \cosh u_1 \frac{\partial V}{\partial u_1} + \frac{1}{\sinh u_1}\frac{\partial^2 V}{\partial u_3{}^2} + \tfrac{1}{4}\sinh u_1 V + \frac{k^2 a^2 \sinh u_1}{\xi^2} V = 0.$$

This equation is only separable when $k = 0$, and for

$$V = V_1(u_1) V_2(u_2) V_3(u_3)$$

we obtain in this case (namely, Laplace's equation)

$$\sinh u_1\left(\frac{1}{V_1}\frac{\mathrm{d}^2 V_1}{\mathrm{d}u_1{}^2} + \frac{1}{V_2}\frac{\mathrm{d}^2 V_2}{\mathrm{d}u_2{}^2}\right) + \cosh u_1 \frac{1}{V_1}\frac{\mathrm{d}V_1}{\mathrm{d}u_1} + \frac{1}{\sinh u_1}\frac{1}{V_3}\frac{\mathrm{d}^2 V_3}{\mathrm{d}u_3{}^2} + \tfrac{1}{4}\sinh u_1 = 0.$$

The substitution $v_1 = \cosh u_1$ now yields the associated Legendre equation

$$\frac{\mathrm{d}}{\mathrm{d}v_1}\left\{(1 - v_1{}^2)\frac{\mathrm{d}V_1}{\mathrm{d}v_1}\right\} + \left(\frac{c_3}{1 - v_1{}^2} - c_2 - \tfrac{1}{4}\right)V_1 = 0.$$

APPENDIX 2

The Schrödinger Equation

The time-independent Schrödinger equation is

$$\frac{\partial^2\psi}{\partial x^2}+\frac{\partial^2\psi}{\partial y^2}+\frac{\partial^2\psi}{\partial z^2}+\frac{2m}{\hbar^2}(E-V)\psi=0,$$

where M, $\hbar$, E are constants and the potential energy V is a function of x, y and z.

(i) LINEAR OSCILLATOR

The reader is referred to section 11.4 for a detailed discussion of the one-dimensional linear oscillator.

(ii) SPHERICALLY SYMMETRIC SYSTEM

In this case V is a function of r when the coordinates are spherical polar. Corresponding to coordinates r, θ, ϕ, the method of separation of variables gives

$$\frac{1}{R}\left[\frac{\mathrm{d}}{\mathrm{d}r}\left(r^2\frac{\mathrm{d}R}{\mathrm{d}r}\right)+\frac{2M}{\hbar^2}\{E-V(r)\}r^2\right]+\frac{1}{P\sin\theta}\frac{\mathrm{d}}{\mathrm{d}\theta}\left(\sin\theta\frac{\mathrm{d}P}{\mathrm{d}\theta}\right)+\frac{1}{\Phi\sin^2\theta}\frac{\mathrm{d}^2\Phi}{\mathrm{d}\phi^2}=0,$$

where R, P and Φ are functions respectively of r, θ and ϕ.

The Φ equation in all cases is simple harmonic, the choice of positive $(\mathrm{d}^2\Phi/\mathrm{d}\phi^2)/\Phi$ being ruled out on physical grounds.

P always satisfies the associated Legendre equation

$$\frac{\mathrm{d}}{\mathrm{d}\mu}\left\{(1-\mu^2)\frac{\mathrm{d}P}{\mathrm{d}\mu}\right\}+\left\{\lambda-\frac{m^2}{1-\mu^2}\right\}P=0,$$

where $\mu=\cos\theta$. The physical conditions yield $\lambda=n(n+1)$, where n is a non-negative integer.

The corresponding equation for R is

$$\frac{\mathrm{d}}{\mathrm{d}r}\left(r^2\frac{\mathrm{d}R}{\mathrm{d}r}\right)+\left[\frac{2M}{h^2}\{E-V(r)\}r^2-n(n+1)\right]R=0$$

and the following cases occur.

(a) *Free particle*

This corresponds to $V = 0$. With $k^2 = 2ME/\hbar^2$ and the substitutions $kr = x$, $R = yx^{-1/2}$, we obtain

$$x^2y'' + xy' + \{x^2 - (n + \tfrac{1}{2})^2\}y = 0,$$

which is Bessel's equation of order $n + \frac{1}{2}$.

(b) *Three-dimensional harmonic oscillator*

This corresponds to $V = \frac{1}{2}kr^2$. With $E = \frac{1}{2}\hbar\omega\lambda$, $k = M\omega$, $r = \sqrt{(\hbar/M\omega)}\bar{r}$ and the substitutions $x = \bar{r}^2$, $R = \bar{r}^n \exp(-\frac{1}{2}\bar{r}^2)y$, we have

$$xy'' + (n + \tfrac{3}{2} - x)y' + \tfrac{1}{4}(\lambda - 3 - 2n)y = 0.$$

The physical requirements yield $\lambda - 3 - 2n = 4k$, where k is a non-negative integer, and we obtain Laguerre's equation

$$xy'' + (n + \tfrac{3}{2} - x)y' + ky = 0.$$

(c) *Kepler problem*

This corresponds to $V = -Ze^2/r$. With $A^2 = -8ME/\hbar^2$ (since $E < 0$ for a bound state), $B = 2MZe^2/A\hbar^2$ $(A > 0)$ and the substitutions $x = Ar$ and $R = y \exp(-\frac{1}{2}x)$, we obtain Laguerre's equation

$$xy'' + (2n + 2 - x)y' + (B - n - 1)y = 0.$$

The physical requirements are that $B - n - 1$ be a non-negative integer.

(iii)

The Kepler problem can also be solved in parabolic coordinates. From Appendix 1 (iv) the Schrödinger equation becomes, in the notation of the last paragraph of that section,

$$\frac{1}{U_1}\frac{\mathrm{d}}{\mathrm{d}u_1}\left(u_1u_2\frac{\mathrm{d}U_1}{\mathrm{d}u_1}\right) + \frac{1}{U_2}\frac{\mathrm{d}}{\mathrm{d}u_2}\left(u_1u_2\frac{\mathrm{d}U_2}{\mathrm{d}u_2}\right) + \frac{1}{U_3}\frac{\mathrm{d}}{\mathrm{d}u_3}\left(\frac{u_1^2 + u_2^2}{u_1u_2}\frac{\mathrm{d}U_3}{\mathrm{d}u_3}\right)$$
$$+ \frac{2M}{\hbar^2}u_1u_2(u_1^2 + u_2^2)\left(E + \frac{2ze^2}{u_1^2 + u_2^2}\right) = 0.$$

The U_3 equation is simple harmonic, and we derive

$$\frac{1}{u_1U_1}\frac{\mathrm{d}}{\mathrm{d}u_1}\left(u_1\frac{\mathrm{d}U_1}{\mathrm{d}u_1}\right) - \frac{n^2}{u_1^2} + \frac{2ME}{\hbar^2}u_1^2 + c_1 = 0,$$
$$\frac{1}{u_2U_2}\frac{\mathrm{d}}{\mathrm{d}u_2}\left(u_2\frac{\mathrm{d}U_2}{\mathrm{d}u_2}\right) - \frac{n^2}{u_2^2} + \frac{2mE}{\hbar^2}u_2^2 + c_2 = 0,$$

where $c_1 + c_2 = 4mZe^2/\hbar^2$ and n is a non-negative integer.

Consider the equation

$$u\frac{\mathrm{d}}{\mathrm{d}u}\left(u\frac{\mathrm{d}U}{\mathrm{d}u}\right)+\left\{-n^2+\frac{2ME}{\hbar^2}u^4+cu^2\right\}U=0.$$

With $A^2 = -2mE/\hbar^2$ and $B = C/A$ and the substitutions $x = Au^2$, $U = \exp(-\frac{1}{2}x)x^{n/2}$, we have Laguerre's differential equation

$$xy''+(n+1-x)y'+\left(\frac{B}{4}-\frac{n+1}{2}\right)y=0.$$

The physical conditions require that $B/4 - (n+1)/2$ be a non-negative integer.

REFERENCES

[1] Arscott, F. M. *Periodic Differential Equations*, Pergamon, Oxford (1964).

[2] Bailey, W. N. *Generalised Hypergeometric Series*, Cambridge Tracts in Mathematics and Mathematical Physics, No. 32, Cambridge University Press, London (1963).

[3] Brillouin, L. *J. Phys. Radium* **7,** 353–368 (1926).

[4] Coddington, E. A. and Levinson, N. *Theory of Ordinary Differential Equations*, McGraw-Hill, New York (1955).

[5] Copson, E. T. *An Introduction to the Theory of Functions of a Complex Variable*, Clarendon Press, Oxford (1935).

[6] Courant, R. and Hilbert, D. *Methods of Mathematical Physics*, Vol. 1, Interscience, New York (1953).

[7] Ince, E. L. *Ordinary Differential Equations*, Dover, New York (1956).

[8] Jeffreys, H. *Proc. Lond. Math. Soc.* (2) **23,** 428–436 (1924).

[9] Kramers, H. A. *Z. Phys.* **39,** 828–840 (1926).

[10] Langer, R. E. *Trans. Am. math. Soc.* **33,** 23–64 (1931).

[11] Langer, R. E. *Trans. Am. math. Soc.* **34,** 447–480 (1932).

[12] Langer, R. E. *Trans. Am. math. Soc.* **36,** 90–106 (1934).

[13] Langer, R. E. *Bull. Am. math. Soc.* **40,** 545–582 (1934).

[14] Spain, B. *Vector Analysis*, Van Nostrand, London (1965).

[15] Titchmarsh, E. C. *Eigenfunction Expansions Associated with Second Order Differential Equations*, Vol. 1, Clarendon Press, Oxford (1962).

[16] Wasow, W. *Asymptotic Expansions for Ordinary Differential Equations*, Interscience, New York (1965).

[17] Wentzel, G. *Z. Phys.* **38,** 518–529 (1926).

[18] Whittaker, E. T. and Watson, G. N. *Modern Analysis* (4th edn), Cambridge University Press, London (1952).

Index